The ASQ Certified Quality Engineer Study Guide

The ASQ Certified Quality Engineer Study Guide

Second Edition

Scott A. Laman, Editor

ASQExcellence
Milwaukee, Wisconsin

Published by ASQExcellence, Milwaukee, WI
Produced and distributed by Quality Press, ASQ, Milwaukee, WI

Publisher's Cataloging-in-Publication Data

Names: Laman, Scott A., author.
Title: The ASQ certified quality engineer study guide , second edition / Scott A. Laman.
Description: Milwaukee, WI: ASQExcellence, 2023.
Identifiers: LCCN: 2023944347 | ISBN: 978-1-63694-113-4 | 978-1-63694-114-1 (pdf) | 978-1-63694-115-8 (epub)
Subjects: LCSH Production management--Quality control--Study guides. | Reliability (Engineering)--Study guides. | BISAC BUSINESS & ECONOMICS / Quality Control | STUDY AIDS / Professional
Classification: LCC TS156 .A87 2023 | DDC 658.4/013--dc23

ASQ and ASQExcellence advance individual, organizational, and community excellence worldwide through learning, quality improvement, and knowledge exchange.

Bookstores, wholesalers, schools, libraries, businesses, and organizations: Quality Press and ASQExcellence books are available at quantity discounts for bulk purchases for business, trade, or educational uses. For more information, please contact Quality Press at 800-248-1946 or books@asq.org.

To place orders or browse the selection of all Quality Press and ASQExcellence titles, visit our website at: http://www.asq.org/quality-press.

Printed in the United States of America.

27 26 GP 7 6 5 4 3 2

Quality Press
600 N. Plankinton Ave.
Milwaukee, WI 53203-2914
Email: books@asq.org
ASQ Excellence Through Quality™

ASQE™ Inspire Excellence

Table of Contents

Foreword

My name is John Lee, and I have passed all 18 of The American Society for Quality (ASQ) certification exams. In the last 25 years, I have also helped prepare approximately 5,000 students around the globe for ASQ certification.

I absolutely recommend *The ASQ Certified Quality Engineer Study Guide, Second Edition* to help you prepare for the ASQ CQE certification exam.

To get the most out of this publication, you should understand some of the more basic elements of effective test taking. I have listed some of them below.

1. Organize your materials. I have taken so many ASQ exams that I have a very clear understanding of what content is tested, and how it is tested.

   A. I recommend tabbing the pages with sticky notes, in *The ASQ Certified Quality Engineer Handbook,* that you feel, with 100% confidence, will show up in the certification exam. I refer to these pages as the "critical stack."

   B. I also recommend making a copy of the index and putting it in a small, separate three-ring binder. This enables you to find important subjects quickly without always having to flip to the back of the book.

2. Know the key quality themes embraced by ASQ. ASQ will test your knowledge of widely accepted best practices and quality principles identified in quality management system standards. In addition, the philosophies of quality gurus such as Deming, Juran, Crosby, and Shewhart will be tested.
   Here is a partial list of priorities that you need to be aware of:

   A. The importance of top management support

   B. Customer focus

   C. Involvement of people

   D. Seeing problems as system problems, not people problems

   E. Continuous improvement, including the PDCA cycle

   F. Factual approach to decision making

   G. Fixing quality problems in design

H. Understanding that preventing problems is desirable over correcting problems

3. Understand test taking logic. A great test taker, in my opinion, will first break a test question down logically. Test taking logic includes such rules as "if an answer choice contains an absolute, then it is likely incorrect." Absolutes include words such as "always" or "never." Word alignment would be another element of test taking logic. Look for clues in how the question is worded that lead you to the answer. The question and answer will likely not contain the exact same key word, but often synonyms are used.
4. Take your test in the proper order. I teach my students to take their test in the following order:

   A. Begin with logic questions. These are questions where you use the rules of logic to arrive at the correct answer.

   B. Look up questions. These are the questions you cannot answer using the rules of logic. This does not include answers with numerical answers.

   C. Number-crunching questions require answers as numbers. These are the most time-consuming and difficult questions. You will need a calculator to answer these questions.
5. Never let time become your constraint. By effectively using the strategy described above, you should finish the exam with plenty of time left over. If you do not follow the strategy effectively, then you may feel you are behind on time. This can easily be converted into test anxiety. Anxiety strips confidence and generally lowers test performance. One of the goals of the strategy described above is to help you maintain your confidence throughout the majority of the test. If you do experience anxiety, then let it take place at the end of the exam. (This is one reason we save the number-crunching questions for last.) Losing confidence at the end of the exam is not nearly as destructive as losing it at the beginning of the exam.

Enjoy your ASQ certification journey!

John Lee
President of Alpha Training and Consulting
Shingo Award Winning Author
BSME, MBA, CMQ/OE, CQE, CRE, CQA, CQIA, CQI, CCT, CQPA, CQT, CHA, CBA, CSQE, CSQP, CSSYB, CSSGB, CSSBB, CPGP, CMBB

Introduction

This book is primarily meant to aid those taking the American Society for Quality (ASQ) Certified Quality Engineer (CQE) exam and is best used in conjunction with *The ASQ Certified Quality Engineer Handbook, Fifth Edition* (ASQExcellence). Section 1 provides 399 practice questions organized by the seven parts of the 2022 Body of Knowledge (BoK). Section 2 gives the reader 216 additional practice questions from each of the seven parts, in a randomized order.

For every question in both sections, solutions are provided that explain why each answer is the correct one and which section of the BoK the question corresponds to so that any further study needed can be focused on specific sections.

In this edition of the Study Guide, many new questions have been written to address the main areas to which the CQE BoK was expanded in 2022. This includes:

- cost-benefit analysis and the responsible, accountable, consulted, and informed matrix (RACI)
- assessing risks in audit planning and implementation
- critical-to-quality concepts
- hazard analysis and use of failure mode and effects analysis
- overall equipment effectiveness (OEE)
- 5 Whys
- data automation and database integration
- risk management

Existing questions have been rewritten so that all are consistent with the types of questions that appear on ASQ certification exams. That is, the questions in this edition:

- are all multiple choice, not true/false.
- do not contain options for answers such as "none of the above" or "all of the above."
- are written from a positive point of view, meaning you will be asked directly which answer <u>is</u> correct, not which answer is <u>not</u> correct.
- have four options for answers to choose from.

- have numerical answers as options in sequence from smallest to largest.
- have been updated as needed to reflect current content, including changes to standards and regulations and the ASQ Code of Ethics.

Questions that were not covered by any of the 2022 BoK areas were deleted. Errata identified by users of the previous edition were addressed. As a result, over half of the content of this study guide is new and improved.

Acknowledgments

The contributors are an illustrious group who also contributed to *The ASQ Certified Quality Engineer Handbook, Fifth Edition*:

- Matthew A. Barsalou, BorgWarner Systems Engineering GmbH
- Andrew Davison, Genesys Spine
- Denis Devos, Devos Associates Inc
- Harish Jose, Bausch + Lomb Synergetics
- Dr. Michael Mladjenovic, The Sensei Group
- Jayet Moon, Terumo Medical Corporation

In addition, I would like to thank Manav Sheth for creating the mathematical formulas and equations for this book.

Scott A. Laman

Certified Quality Engineer (CQE) Body of Knowledge

The topics in this Body of Knowledge include subtext explanations and the cognitive level at which the questions will be written. This information will provide useful guidance for both the Exam Development Committee and the candidate preparing to take the exam. The subtext is not intended to limit the subject matter or be all-inclusive of that material that will be covered in the exam. It is meant to clarify the type of content that will be included on the exam. The descriptor in parentheses at the end of each entry refers to the maximum cognitive level at which the topic will be tested. A complete description of cognitive levels is provided at the end of this document.

I. Management and Leadership (17 Questions)

A. Quality Philosophies and Foundations

Describe continuous improvement tools, including lean, six sigma, statistical process control (SPC), and total quality management. Understand how modern quality has evolved from quality control through statistical process control (SPC) to total quality management and leadership principles (including Deming's 14 points). (Understand)

B. The Quality Management System (QMS)

1. Strategic planning

Identify and define top management's responsibility for the QMS, including establishing policies and objectives, setting organization-wide goals, and supporting quality initiatives. (Apply)

2. Deployment techniques

Define, describe, and use various deployment tools in support of the QMS such as:

a. Benchmarking

Define the concept of benchmarking and why it may be used. (Remember)

b. Stakeholder

Define, describe, and use stakeholder identification and analysis. (Apply)

c. Performance

Define, describe, and use performance measurement tools such as cost-benefit analysis. (Apply)

d. Project management

Define, describe, and use project management tools, including Gantt charts and the responsible, accountable, consulted and informed matrix (RACI). (Apply)

3. Quality information system (QIS)

Identify and describe the basic elements of a QIS, including who will contribute data, the kind of data to be managed, who will have access to the data, the level of flexibility for future information needs, and data analysis. (Understand)

C. ASQ Code of Ethics for Professional Conduct

Determine appropriate behavior in situations requiring ethical decisions. (Evaluate)

D. Leadership Principles and Techniques

Analyze various principles and techniques for developing and organizing teams and leading quality initiatives. (Analyze)

E. Facilitation Principles and Techniques

1. Roles and responsibilities

Describe the facilitator's roles and responsibilities on a team. (Understand)

2. Facilitation tools

Apply various tools used with teams, including brainstorming, nominal group technique, conflict resolution, and force-field analysis. (Apply)

F. Communication Skills

Identify and distinguish between specific communication methods that are used for delivering information and messages in a variety of situations across all levels of the organization. (Analyze)

G. Customer Relations

Define, apply, and analyze the results of customer relation tools such as customer satisfaction surveys. (Analyze)

H. Supplier Management

1. Techniques

Apply various supplier management techniques, including supplier qualification, certification, and evaluation. (Apply)

2. Improvement

Analyze supplier ratings and performance improvement results. (Analyze)

3. Risk

Understand business continuity, resiliency, and contingency planning. (Understand)

I. Barriers to Quality Improvement
Identify barriers to quality improvement, analyze their causes and impact, and implement methods for improvement. (Analyze)

II. The Quality System (18 Questions)

A. Elements of the Quality System

1. Basic elements
Interpret the basic elements of a quality system, including planning, control, and improvement, from product and process design through quality cost systems and audit programs. (Evaluate)

2. Design
Analyze the design and alignment of interrelated processes to the strategic plan and core processes. (Analyze)

B. Documentation of the Quality System

1. Document components
Identify and describe quality system documentation components, including quality policies and procedures to support the system. (Understand)

2. Document control
Evaluate configuration management, maintenance, and document control to manage work instructions and quality records. (Evaluate)

C. Quality Standards and Other Guidelines
Apply national and international standards and other requirements and guidelines, including the Malcolm Baldrige National Quality Award (MBNQA), and describe key points of the ISO 9000 series of standards. [Note: Industry-specific standards will not be tested.] (Apply)

D. Quality Audits

1. Types of audits
Describe and classify the various types of quality audits such as product, process, management (system), registration (certification), compliance (regulatory), first, second, and third party. (Apply)

2. Roles and responsibilities in audits
Identify and define roles and responsibilities for audit participants such as audit team (leader and members), client, and auditee. (Understand)

3. Audit planning and implementation
Describe and apply the stages of a quality audit, from audit planning, including assessing risks through conducting an audit. (Apply)

4. **Audit reporting and follow-up**
   Apply the steps of audit reporting and follow up, including the need to verify corrective action. (Apply)

E. **Cost of Quality (COQ)**
Identify and apply COQ concepts, including cost categorization, data collection, reporting, and interpreting results. (Analyze)

F. **Quality Training**
Identify and apply key elements of a training program, including conducting a needs analysis, preparing curricula and materials, and determining the program's effectiveness. (Apply)

III. Product, Process, and Service Design (21 Questions)

A. **Classification of Quality Characteristics**
Define, interpret, assess, and classify quality characteristics for new and existing products, processes, and services. [Note: The classification of defects is covered in IV.B.3.] (Evaluate)

B. **Design Inputs, Techniques, and Review**

1. **Inputs**
   Classify design inputs such as customer needs, regulatory requirements, critical to quality, and risk assessment into robust design using techniques such as failure mode and effects analysis (FMEA). (Analyze)

2. **Techniques**
   Apply Design for X (DFX), Design for Six Sigma (DFSS), and requirements traceability. (Apply)

3. **Review**
   Identify and apply common elements of the design review process, including roles and responsibilities of participants. (Apply)

C. **Technical Drawings and Specifications**
Interpret specification requirements in relation to product and process characteristics and technical drawings, including characteristics such as views, title blocks, dimensioning and tolerancing, and geometric dimensioning and tolerance symbols (GD&T). (Evaluate)

D. **Verification and Validation**
Interpret the results of evaluations and tests used to verify and validate the design of products, processes and services, such as installation qualification (IQ), operational qualification (OQ), and process qualification (PQ). (Evaluate)

E. Reliability and Maintainability

1. Predictive and preventive maintenance tools
Describe and apply the tools and techniques used to maintain and improve process and product reliability. (Apply)

2. Reliability and maintainability indices
Apply indices such as mean time to failure (MTTF), mean time between failure (MTBF), mean time to repair (MTTR), availability, and failure rate. (Apply)

3. Reliability models
Identify, define, and distinguish between the basic elements of reliability models such as exponential, Weibull, and bathtub curve. (Apply)

4. Reliability/safety/hazard assessment tools
Define, construct, and interpret the results of failure mode and effects analysis (FMEA), design FMEA (dFMEA), process FMEA (pFMEA), use FMEA (uFMEA), failure mode, effects, and criticality analysis (FMECA), and hazard analysis. (Analyze)

IV. Product and Process Control (23 Questions)

A. Methods
Implement product and process control methods such as control plan development, critical control point identification, and work instruction development and validation. (Analyze)

B. Material Control

1. Material identification, status, and traceability
Define and distinguish between these concepts, and describe methods for applying them in various situations. (Analyze)

2. Material segregation
Describe material segregation and its importance, and evaluate appropriate methods for applying it in various situations. (Evaluate)

3. Material classification
Assess and classify product and process defects and non-conformities. (Evaluate)

4. Material review board (MRB)
Describe the purpose and function of an MRB and evaluate nonconforming product or material to make a disposition decision in various situations. (Evaluate)

C. **Acceptance Sampling**

1. **Sampling concepts**
   Apply the concepts of producer and consumer risk, and related terms, including operating characteristic (OC) curves, acceptable quality limit (AQL), and lot tolerance percent defective (LTPD). (Apply)

2. **Sampling standards and plans**
   Identify, interpret, and apply ANSI/ASQ Z1.4 and Z1.9 standards for attributes and variables sampling. (Analyze)

3. **Sample integrity**
   Identify and apply techniques for establishing and maintaining sample integrity. (Apply)

D. **Measurement and Test**

1. **Measurement tools**
   Select and describe appropriate uses of inspection tools such as gage blocks, calipers, micrometers, optical comparators, and coordinate measuring machines (CMM). (Analyze)

2. **Destructive and nondestructive tests**
   Identify when destructive and nondestructive measurement test methods should be used and apply the methods appropriately. (Apply)

E. **Metrology**
Apply metrology techniques such as calibration, traceability to calibration standards, measurement error and its sources, and control and maintenance of measurement standards and devices. (Apply)

F. **Measurement System Analysis (MSA)**
Calculate, analyze, and interpret repeatability and reproducibility (Gage R&R) studies, measurement correlation, capability, bias, linearity, precision, stability and accuracy, using MSA quantitative and graphical methods. (Evaluate)

V. **Continuous Improvement (26 Questions)**

A. **Quality Control Tools**
Select, construct, apply, and interpret the following quality control tools:

1. Flowcharts
2. Pareto charts
3. Cause and effect diagrams
4. Control charts

5. Check sheets
6. Scatter diagrams
7. Histograms (Analyze)

B. Quality Management and Planning Tools
Select, construct, apply, and interpret the following quality management and planning tools:

1. Affinity diagrams and force field analysis
2. Tree diagrams
3. Process decision program charts (PDPC)
4. Matrix diagrams
5. Interrelationship digraphs
6. Prioritization matrices
7. Activity network diagrams (Analyze)

C. Continuous Improvement Methodologies
Define, describe, and apply the following continuous improvement methodologies:

1. Total quality management (TQM)
2. Kaizen
3. Plan-do-check-act (PDCA)
4. Six sigma (Analyze)

D. Lean tools
Define, describe, and apply the following lean tools:

1. 5S
2. Value-stream mapping
3. Kanban
4. Visual control
5. 8 Wastes
6. Standardized work
7. Takt time

8. Single minute exchange of die (SMED)

9. Overall equipment effectiveness (OEE) (Evaluate)

E. Corrective Action
Identify, describe, and apply elements of the corrective action process, including problem identification, failure analysis, root cause analysis, 5 Whys, problem correction, recurrence control, and verification of effectiveness. (Evaluate)

F. Preventive Action
Identify, describe and apply various preventive action tools such as error-proofing/poka-yoke and robust design and analyze their effectiveness. (Evaluate)

VI. Quantitative Methods and Tools (34 Questions)

A. Collecting and Summarizing Data

1. Types of data
Define, classify, and compare discrete (attributes) and continuous (variables) data. (Apply)

2. Measurement scales
Define and describe nominal, ordinal, interval, and ratio scales. (Understand)

3. Data collection methods
Describe various methods for collecting data, including tally or check sheets, data coding, automatic gaging, data automation, database integration, and identify the strengths and weaknesses of the methods. (Apply)

4. Data accuracy and integrity
Identify factors that can influence data accuracy such as source/resource issues, flexibility, versatility, inconsistency, inappropriate interpretation of data values, and redundancy to ensure data accuracy and integrity. (Apply)

5. Data visualization techniques
Apply and interpret data visualization techniques using dashboards, and select the appropriate metrics for dashboards. (Apply)

6. Descriptive statistics
Describe, calculate, and interpret measures of central tendency and dispersion, apply the central limit theorem, and construct and interpret frequency distributions, including simple, categorical, grouped, ungrouped, and cumulative. (Evaluate)

7. Graphical methods for depicting distributions
Apply and interpret diagrams such as probability plots for normal and other distributions. [Note: Histograms are cov ered in V.A.] (Analyze)

B. Quantitative Concepts

1. **Terminology**
   Define and apply quantitative terms, including population, parameter, sample, statistic, random sampling, and expected value. (Analyze)

2. **Drawing statistical conclusions**
   Distinguish between numeric and analytical studies. Assess the validity of statistical conclusions by analyzing the assumptions used and the robustness of the technique used. (Evaluate)

3. **Probability terms and concepts**
   Describe concepts such as independence, mutual exclusivity, multiplication rules, complementary probability, and joint occurrence of events. (Apply)

C. Probability Distributions

1. **Continuous distributions**
   Define and distinguish between these distributions such as normal, uniform, exponential, lognormal, Weibull, Student's *t* and *F*. (Analyze)

2. **Discrete distributions**
   Define and distinguish between these distributions such as binomial, Poisson, hypergeometric, and multinomial. (Analyze)

D. Statistical Decision-Making

1. **Point estimates and confidence intervals**
   Define, describe, and assess the bias of estimators. Calculate and interpret standard error, tolerance intervals, and confidence intervals. (Evaluate)

2. **Hypothesis testing**
   Define, interpret, and apply hypothesis tests for means, variances, and proportions. Apply and interpret the concepts of significance level, power, type I, and type II errors. Define and distinguish between statis- tical and practical significance. (Evaluate)

3. **Paired-comparison tests**
   Define and use paired-comparison (parametric) hypothesis tests, and interpret the results. (Apply)

4. **Goodness-of-fit tests**
   Define and use chi square and other goodness-of-fit tests, and understand the results. (Apply)

5. **Analysis of variance (ANOVA)**
   Define use, and interpret ANOVA and interpret the results. (Analyze)

6. **Contingency tables**
   Define and use contingency tables to evaluate statistical significance. (Apply)

E. **Relationships Between Variables**

1. **Linear regression**
   Calculate simple linear regression models. Illustrate hypothesis tests for regression statistics. Use linear regression models for estimation and prediction. (Apply)

2. **Simple linear correlation**
   Calculate the correlation coefficient and its confidence interval, and illustrate a hypothesis test for correlation statistics. (Apply)

3. **Time-series analysis**
   Define, describe, and use time-series analysis, including moving aver- age to identify trends and seasonal or cyclical variation. (Apply)

F. **Statistical Process Control (SPC)**

1. **Objectives and benefits**
   Identify and explain the objectives and benefits of SPC. (Understand)

2. **Common and special causes**
   Describe, identify, and distinguish between these types of causes. (Analyze)

3. **Selection of variable**
   Identify and select variable characteristics for monitoring by control chart. (Analyze)

4. **Rational subgrouping**
   Define and apply the principles of rational subgrouping. (Apply)

5. **Control charts**
   Identify, select, construct, and use various control charts, including $\bar{x}$ - R, $\bar{x}$ - s, individuals and moving range (ImR or XmR), moving average and moving range (MAMR), p, np, c, and u. (Analyze)

6. **Control chart analysis**
   Read and interpret control charts and use rules for determining statis- tical control. (Evaluate)

7. **Short-run SPC**
   Identify and define short-run SPC rules. (Understand)

G. Process and Performance Capability

1. Process capability studies

Define, describe, calculate, and use process capability studies, including identifying characteristics, specifications and tolerances, developing sampling plans for such studies, and establishing statistical control. (Analyze)

2. Process performance vs. specifications

Distinguish between natural process limits and specification limits, and calculate percent defective, defects per million opportunities (DPMO), and parts per million (ppm). (Analyze)

3. Process capability indices

Define, select, and calculate C_p , C_{pk}, C_{pm}, and C_r, and evaluate process capability. (Evaluate)

4. Process performance indices

Define, select, and calculate P_p and P_{pk}, and evaluate process performance. (Evaluate)

H. Design and Analysis of Experiments

1. Terminology

Define terms such as dependent and independent variables, factors, levels, response, treatment, error, and replication. (Understand)

2. Planning and organizing experiments

Identify the basic elements of designed experiments, including deter- mining the experiment objective, selecting factors, responses, and measurement methods, and choosing the appropriate design. (Analyze)

3. Design principles

Define and apply the principles of power and sample size, balance, replication, order, efficiency, randomization, blocking, interaction, and confounding. (Apply)

4. Full-factorial experiments

Construct full-factorial designs and use computational and graphical methods to analyze the significance of results. (Analyze)

5. Two-level fractional factorial experiments

Construct two-level fractional factorial designs and apply computational and graphical methods to analyze the significance of results. (Analyze)

VII. Risk Management (21 Questions)

A. Risk Fundamentals

1. Risk terminology
Define, describe, and apply risk terminology such as risk, risk management, severity, occurrence, detection, and risk-based thinking. (Analyze)

2. Types of risk management
Understand and apply various types of enterprise (strategic, software, business, regulatory, medical, audit), operational (supplier, supply chain, safety, project, manufacturing, operations, service, quality system), and product (design, process, use, safety) risk management. (Apply)

B. Risk Planning and Assessment

1. Risk management plan
Analyze and interpret a risk management plan and its components (objectives, risk criteria, stakeholder identification, and team member roles/responsibilities) to identify and prioritize risks. (Analyze)

2. Risk assessment
Apply categorization methods and evaluation tools to assess risk such as failure mode and effects analysis. Identify and apply evaluation metrics including the use of risk matrices, risk priority numbers, and acceptability criteria. (Analyze)

C. Risk Treatment, Control, and Monitoring

1. Identification and documentation
Identify risks, gaps, and controls and document with tools such as a risk register. (Analyze)

2. Risk management system evaluation
Apply auditing techniques and testing of controls to evaluate a risk management system. (Apply)

3. Risk treatment strategies
Understand and apply risk treatment strategies, such as avoid, mitigate, transfer, and accept. (Analyze)

4. Risk monitoring
Apply risk monitoring techniques such as, complaint tracking, trending, and post-market surveillance. (Analyze)

5. Mitigation planning
Apply and interpret risk mitigation plan. (Analyze)

LEVELS OF COGNITION BASED ON BLOOM'S TAXONOMY—REVISED (2001)

In addition to content specifics, the subtext for each topic in this BOK also indicates the intended complexity level of the test questions for that topic. These levels are based on "Levels of Cognition" (from Bloom's Taxonomy—Revised, 2001) and are presented below in rank order, from least complex to most complex.

Remember

Recall or recognize terms, definitions, facts, ideas, materials, patterns, sequences, methods, principles.

Understand

Read and understand descriptions, communications, reports, tables, diagrams, directions, regulations.

Apply

Know when and how to use ideas, procedures, methods, formulas, principles, theories.

Analyze

Break down information into its constituent parts and recognize their relationship to one another and how they are organized; identify sublevel factors or salient data from a complex scenario.

Evaluate

Make judgments about the value of proposed ideas, solutions, by comparing the proposal to specific criteria or standards.

Create

Put parts or elements together in such a way as to reveal a pattern or structure not clearly there before; identify which data or information from a complex set is appropriate to examine further or from which supported conclusions can be drawn.

Section 1

Sample Questions by BoK

Part I	Management and Leadership
Part II	The Quality System
Part III	Product, Process, and Service Design
Part IV	Product and Process Control
Part V	Continuous Improvement
Part VI	Quantitative Methods and Tools
Part VII	Risk Management

Section 1 is divided into seven parts, one for each section in the Certified Quality Engineer Body of Knowledge. In each part, there is a set of questions followed by detailed solutions.

Part I
Management and Leadership

(33 questions)

QUESTIONS

1. Quality planning, quality control, and quality improvement are parts of the quality trilogy attributed to:

   a. Deming.

   b. Crosby.

   c. Juran.

   d. Ishikawa.

2. According to the quality trilogy, quality improvement is:

   a. monitoring techniques to correct sporadic problems.

   b. an annual quality program to institutionalize managerial control and review.

   c. a method that drives out fear so that everyone may work effectively.

   d. a breakthrough sequence to solve chronic problems.

3. Which of the following is one of Deming's 14 points?

   a. Eliminate work standards that prescribe numerical quotas.

   b. Institute barriers that stand between the hourly worker and his right to pride of workmanship.

   c. 50% of quality problems are due to employees.

   d. 15% of quality problems are due to management.

4. Quality defined as "fitness for use" is attributed to:

   a. Deming.

   b. Crosby.

c. Feigenbaum.

d. Juran.

5. SWOT is the acronym for:

   a. strengths, weaknesses, outcomes, and threats.

   b. strengths, weaknesses, opportunities, and threats.

   c. suppliers, workers, opportunities, and tables.

   d. strengths, weaknesses, opportunities, and tables.

6. Which of the following would be considered an effectiveness test for strategic planning?

   a. Have modern methods of training on the job been instituted?

   b. Have only those people who have experience in a particular job position been hired?

   c. Are the goals and objectives aligned throughout the organization?

   d. Have unclear slogans been eliminated?

7. In strategic planning, SMART represents:

   a. strengths, machines, applications, realistic, and time-based.

   b. strengths, measurable, achievable, realistic, and threats.

   c. specific, measurable, achievable, realistic, and time-based.

   d. specific, measurable, applications, reliability, and time-based.

8. Action plans as part of the goals and objectives of a deployment technique are:

   a. to continually build and retain a loyal customer base.

   b. to deliver all products to all customers 100% on time.

   c. detailed plans stating how, when, and by whom the objective will be achieved.

   d. undocumented and unnecessary to monitor.

9. The process organizations use to evaluate their performance against their competition or best practices found internally is referred to as:

   a. benchmarking.

   b. CPM.

c. deployment.

d. stakeholder analysis.

10. Benchmarking projects will include the following steps:

    a. Planning, detection, analysis, and implementation.

    b. Planning, data collection, analysis, and implementation.

    c. Promotion, detection, analysis, and implementation.

    d. Promotion, data collection, analysis, and planning.

11. The owners of a company are the:

    a. stockholders.

    b. executive group.

    c. suppliers.

    d. customers.

12. Business processes, customers, and financial fundamentals are all perspectives of what philosophy?

    a. Statistical process control

    b. Deming's 14 points

    c. Balanced scorecard

    d. RACI

13. Performance measures:

    a. should be achievable.

    b. do not need to be measurable.

    c. do not need to be linked to strategic objectives.

    d. should focus on as many measures as possible.

14. The premise of cost-benefit analysis is that:

    a. things that cost more have more benefit.

    b. investing for the longer term is the best strategy.

    c. short-term pain can increase long term gain.

    d. investments can be compared based on financial analysis.

15. To consider the costs and benefits of a future project it is necessary to:

    a. express the value of present-day dollars as some projected future value.

    b. express future costs and benefits in present-day dollars.

    c. know what interest rates in the future will be.

    d. know the lifecycle of the product or project being considered.

16. In a RACI Matrix, the letters RACI stand for:

    a. Reportable, Accountable, Constructed, Informed.

    b. Roles, Actions, Considerations, Information.

    c. Responsible, Accountable, Consulted, Informed.

    d. Responsible, Actionable, Consulted, Informed.

17. According to RACI philosophy, how many people can be in the Accountable role for a task?

    a. As many as are appropriate

    b. One

    c. Two or three at the most

    d. The same number as are in the Responsible role

18. Which of the following tools can be used for justifying and prioritizing projects?

    a. Net present value (NPV)

    b. Statistical process control (SPC)

    c. Design of experiments (DOE)

    d. Coordinate measuring machine (CMM)

19. Suppose net present value (NPV) is to be estimated for a particular project. To determine how the NPV will change if the interest rate decreases or increases, for example, by 5%, one is doing a type of:

    a. decision tree analysis.

    b. internal rate of return analysis.

    c. payback period analysis.

    d. sensitivity analysis.

20. A Gantt chart is a useful:

    a. financial tool.

    b. planning tool.

    c. statistical analysis tool.

    d. defect classification tool.

21. Which of the following systems is used to record critical data and control and monitor a process?

    a. Quality standards system

    b. Quality audit system

    c. Cost of quality system

    d. Quality information system

22. The ASQ Code of Ethics primarily regulates its members':

    a. styles.

    b. behaviors.

    c. opinions.

    d. personalities.

23. When citing the ASQ CQE certification on social media, one should make sure that their certification is:

    a. in progress.

    b. current.

    c. expired.

    d. a certificate.

24. The ASQ Code of Ethics contains a statement that approves of:

    a. plagiarism.

    b. social responsibility.

    c. basing decisions on feelings.

    d. disclosing confidential information.

25. Which of the following would be an ASQ Code of Ethics violation?
    a. Filing a patent for your work
    b. Being objective in quality-based decisions
    c. Hiring the best candidate for a job
    d. Providing consulting on an unfamiliar topic
26. The most severe discipline resulting from an ASQ Code of Ethics violation is:
    a. exclusion from leadership.
    b. written warning.
    c. expulsion from ASQ.
    d. temporary revocation of certifications.
27. Common development stages teams often progress through are known as *forming*, *storming*, *norming*, and *performing*. In the "storming" stage:
    a. individuals begin to shift from personal concerns to the needs of the team.
    b. the team's mission is clarified and specific roles are identified.
    c. team members still think and act as individuals instead of impacting as a team.
    d. the team has matured and is working in the best interest of the team and team goals.
28. Keeping the team on task is the main function of the team:
    a. leader.
    b. facilitator.
    c. note taker.
    d. sponsor.
29. Part of the team facilitator's mission is to:
    a. discuss content issues with the team.
    b. discuss process issues with the team.
    c. suggest technical solutions to the team.
    d. provide financial backing to the team.

30. A useful brainstorming tool for reducing a large list of items to a smaller, manageable one is:

    a. quality function deployment.

    b. fault tree analysis.

    c. nominal group technique.

    d. operating characteristic curve.

31. In the communication process, there are five methods of feedback: evaluation, interpretation, support, probing, and understanding. Of these methods, the one that most often creates defensiveness and can quickly break the communication process is:

    a. evaluation.

    b. support.

    c. probing.

    d. understanding.

32. Which of the following supplier management tools is used to initially verify that the supplier has adequate financial resources, manufacturing capabilities, and quality systems?

    a. Supplier audit

    b. Supplier classifications

    c. Supplier survey

    d. Supplier performance improvement

33. Implementing quality improvement in an organization can be a failure for several reasons. Which of the following is one obstacle to implementing quality?

    a. Lack of real employee empowerment

    b. Emphasis on a formal strategic plan for change

    c. Strong workforce motivation

    d. Focus on the customer

SOLUTIONS

1. c; Juran understood that improving quality required different approaches and effort than simply maintaining quality. The Juran trilogy focused on three philosophies: quality planning, quality control, and quality improvement. [I.A.1]

2. d; In his 1964 text, *Managerial Breakthrough*, Juran defined quality improvement as a breakthrough sequence to solve chronic problems that is analogous to "common causes." [I.A.1]

3. a; Deming argued that work standards that include numerical quotas do nothing to motivate the common worker and should be eliminated. [I.A.1]

4. d; Juran believed that a product is fit for use if it meets customer expectations. [I.A.2]

5. b; SWOT analysis (strengths, weaknesses, opportunities, and threats) is conducted to assess strengths and weaknesses of an organization, to predict opportunities, and to identify threats from competitors. [I.B.1]

6. c; a and b are two of Deming's fourteen points, while d does not apply, since Deming wanted to eliminate slogans. Goals and objectives aligned throughout the organization are necessary to carry out the strategies to be put in place. [I.B.1]

7. c; Goals and objectives should be "specific, measurable, achievable, realistic, and time-based" to be effective throughout the enterprise. [I.B.1]

8. c; Action plans often resemble mini project plans, or more-complex project planning documents may be in order. [I.B.2]

9. a; Benchmarking is the process by which companies compare their processes against best practices found either internally or externally. [I.B.2.a]

10. b; Typically, benchmarking projects must have a plan, data must be collected for analysis, and finally, there must be implementation of results of the analysis. [I.B.2.a]

11. a; The stockholders of a company are considered the owners of the company. Their role is of a financial nature, but generally passive. [I.B.2.b]

12. c; Business processes, customers, and financial fundamentals are part of the balanced scorecard, with the other category being learning and growth. [I.B.2.b]

13. a; All performance measures should be achievable and aligned with strategic goals. In addition, the focus should be on the "vital few." One should avoid using too many measures or metrics. [I.B.2.c]

14. d; Cost-benefit analysis deals with comparing potential investments with each other to make the wisest decisions. [I.B.2.c]

15. b; In cost-benefit analysis, the financial analysis is performed based on translations to today's value. [I.B.2.c]

16. c; RACI stand for Responsible, Accountable, Consulted, Informed. [I.B.2.d]

17. b; The Accountable person is ultimately answerable for the successful completion of the task, so there can be only one person where "the buck stops here." [I.B.2.d]

18. a; Net present value is the only option used for justifying and prioritizing projects. [I.B.2.d]

19. d; When metrics are examined by varying the inputs (such as percentage, rate, and so on) in NPV, we are conducting a sensitivity analysis. [I.B.2.d]

20. b; A Gantt chart is also a milestone chart—a planning tool. [I.B.2.d]

21. d; A quality information system is the collection of data, rules, and equipment that creates information about quality in a systematic way. [I.B.3]

22. b; Per the ASQ Code of Ethics Introduction, "The purpose of the American Society for Quality (ASQ) Code of Ethics is to establish global standards of conduct and behavior for its members, certification holders, and anyone else who may represent or be perceived to represent ASQ." [1.C]

23. b; The ASQ Code of Ethics says, "Accurately represent and do not mislead others regarding professional qualifications, including education, titles, affiliations, and certifications." Previous and future certifications do not hold the same credibility as current certifications. [I.C]

24. b; The ASQ Code of Ethics contains a section, "Act and conduct business in a professional and socially responsible manner." The other options would not be condoned. [I.C]

25. d; The ASQ Code of Ethics says, "Offer services, provide advice, and undertake assignments only in your areas of competence, expertise, and training." [I.C]

26. c; The ASQ Code of Ethics says, "Disciplinary actions will be commensurate with the seriousness of the offense and may include permanent revocation of certifications and/or expulsion from the society." [I.C]

27. c; In the storming phase, teams become familiar with one another, have worked on missions and strategies, but often are still working in the individuals' best interests and not necessarily the team's as a whole. [I.D]

28. b; The facilitator's role is to provide support for the team, and the main task for that individual is to keep the team on task. [I.E.1]

29. b; The facilitator's role should be marginal and grow less important as the team matures. Overall, the facilitator is responsible for process issues—meeting agendas, communication among teams, and keeping the team on task. [I.E.1]

30. c; The nominal group technique allows for brainstorming a large number of ideas/suggestions while providing everyone an equal voice in the process. [I.E.2]

31. a; All feedback methods can lead to a breakdown in the communication process. However, evaluation, which involves a judgment being made about the importance, worth, or appropriateness of the statement, is the one that will lead to a breakdown much more quickly. It will also take a great deal of effort to recover from such feedback. [I.F]

32. c; Supplier audit is not a bad option, but supplier survey is better, since it provides an initial assessment. The purpose of the survey of a supplier or potential supplier is to make sure their financial resources as well as their manufacturing capabilities and quality systems are acceptable. These three arenas are surveyed by appropriate groups involved with each activity (manufacturing capability surveyed by manufacturing engineer, for example). [I.H]

33. a; The *ASQ Certified Quality Engineer Handbook*, Fifth Edition, provides an overview of twelve common barriers or obstacles to a successful implementation of quality improvement in an organization: lack of time to devote to quality initiatives, poor intraorganizational communication, lack of real employee empowerment, lack of employee trust in senior management, politics and turf issues, lack of a formalized strategic plan for change, lack of strong motivation, view of quality as a quick fix, drive for short-term financial results, lack of leadership, lack of customer focus, lack of a company-wide definition of quality. [I.I]

Part II
The Quality System

(33 questions)

QUESTIONS

1. For any quality system, all quality-related activities begin with:
   a. sales and distribution.
   b. process planning.
   c. customer needs.
   d. product design.
2. A quality manual generally consists of four *tiers* or *layers* that include:
   a. policies, procedures, instructions, and records.
   b. policies, documentation, instructions, and records.
   c. documentation, implementation, compliance, and accuracy.
   d. compliance, accuracy, clarity, and records.
3. The critical characteristics of any documentation of a quality system are:
   a. compliance, implementation, and clarity.
   b. compliance, clarity, and procedures.
   c. procedures, accuracy, and clarity.
   d. compliance, accuracy, and clarity.
4. One major component of a quality system is:
   a. supplier management.
   b. implementation of a documented quality system.
   c. contractor assessment.
   d. auditing of suppliers.

5. Once a quality manual is complete, a final review is in order. The reviewed copy should be endorsed and authorized for release by:

   a. suppliers.

   b. customers.

   c. top management.

   d. the chief quality officer.

6. Which of the following is a requirement for maintaining document control?

   a. A process is in place for distribution of revisions and removal of obsolete documents.

   b. A process is in place for making the documents readily available to the general public.

   c. Documents are approved by top management.

   d. Hard copies of documents must be maintained.

7. ISO 9001:2015 is:

   a. a national standard.

   b. an international standard.

   c. an industry-specific standard.

   d. a guidance document.

8. Which of the following is the subject of one of the standards in the ISO 9000 Systems family?

   a. Environmental management systems

   b. Continual improvement

   c. Risk management

   d. Quality management

9. The ISO 9000 series of standards emphasizes eight quality management principles. Which of the following would be categorized as one of those principles?

   a. Understanding customer needs

   b. Use of technical language in all technical descriptions

c. Use of different terms and vocabulary

d. Establishing functional silos

10. Which of the following ISO standards are guidelines only?

    a. ISO 9000

    b. ISO 9001

    c. ISO 9004

    d. ISO 14001

11. The Malcolm Baldrige National Quality Award consists of how many key criteria for performance excellence?

    a. 4

    b. 7

    c. 8

    d. 14

12. A quality audit is to be conducted within a particular company. An employee of this company will be conducting the audit. This is an example of:

    a. an internal quality audit.

    b. an external quality audit.

    c. a third-party quality audit.

    d. a second-party audit.

13. The most comprehensive type of audit is the:

    a. quality system audit.

    b. product quality audit.

    c. process quality audit.

    d. service quality audit.

14. Which of the following would be a purpose for an internal quality audit?

    a. To meet requirements for certification to a management standard

    b. Verification of conformance with contractual requirements

    c. To contribute to the improvement of the management system

    d. To maintain confidence in the capability of a supplier

15. A quality audit has been conducted and some corrective measures are recommended. The party responsible for planning and implementing these measures is:

    a. the auditee.

    b. the lead auditor.

    c. management where the corrective measures are needed.

    d. line workers where the corrective measures are needed.

16. Corrective action and verification that nonconformities have been addressed are:

    a. always carried out.

    b. often poorly completed.

    c. unnecessary and cause more stress than it is worth.

    d. performed before root cause analysis.

17. The lead auditor must explain to the auditee the objectives of the audit as well as the methods that will be used. This is usually done during:

    a. the opening meeting (or pre-examination).

    b. the closing meeting.

    c. the audit itself.

    d. no time; it is not necessary for the auditor to explain this information to the auditee.

18. Who is responsible for the accuracy of the final audit report?

    a. Auditee

    b. Client

    c. Lead Auditor

    d. Auditor

19. When assessing Audit nonconformities as an Auditor, it is recommended to use a(n):

    a. audit checklist.

    b. risk-based approach.

    c. statistical technique approach.

    d. third-party inspection procedure.

20. Audit planning by the lead Auditor should include consideration of:
    a. the auditee, auditor, and closing meeting.
    b. the management review meeting, final inspection, and design reviews.
    c. the audit scope, depth, and time frame.
    d. the ISO certification body audit, FDA inspection, and OSHA audit.
21. The auditee will begin improvement efforts to address any problems or corrections identified by the audit. A follow-up audit to verify results of this improvement effort:
    a. should be agreed on by all parties: client, auditor, auditee.
    b. is not necessary or recommended.
    c. is conducted by someone in the company when there is time.
    d. is communicated via memo from the auditee.
22. A company increased prevention costs by making changes to its quality control area, and then nonconformities decreased. As a result, there was a reduction in:
    a. operating costs.
    b. appraisal costs.
    c. failure costs.
    d. quality costs.
23. Correction of data entry errors, specimen processing, and cost of reeducating staff in a laboratory before delivery of the results to the customer are examples of:
    a. prevention costs.
    b. appraisal costs.
    c. internal failure costs.
    d. external failure costs.
24. In one year, the cost of proficiency testing for quality assurance inspectors by an external certified agency for all employees was $10,000. This cost would be categorized as:
    a. prevention cost.
    b. appraisal cost.
    c. internal failure cost.
    d. external failure cost.

25. Money spent on review of suppliers is:

    a. prevention cost.

    b. appraisal cost.

    c. external failure cost.

    d. internal failure cost.

26. As failure costs decrease due to improved quality, what typically happens to appraisal activities?

    a. Appraisal activities increase.

    b. Appraisal activities decrease.

    c. There is no relationship between the two.

    d. Appraisal activities will remain the same.

27. Test equipment depreciation is an example of:

    a. prevention cost.

    b. appraisal cost.

    c. external failure cost.

    d. internal failure cost.

28. Last month, a local company reported the following quality costs.

Training	$40,000
Material inspection	$12,000
Test equipment	$65,000
In-plant scrap and rework	$15,000

    What is the total appraisal cost for last month?

    a. $12,000

    b. $65,000

    c. $77,000

    d. $92,000

29. Last month, a local company reported the following quality costs.

Training	$40,000
Material inspection	$12,000
Test equipment	$65,000
In-plant scrap and rework	$15,000

What is the reported external failure cost for last month?

a. $0

b. $12,000

c. $77,000

d. $92,000

30. A quality system audit is to be conducted. This is an example of a(n):

a. prevention cost.

b. appraisal cost.

c. external failure cost.

d. internal failure cost.

31. The first step in developing a training program is:

a. design and development of the curriculum, or training plan.

b. developing the lesson plan.

c. conducting a pretest for the participants.

d. assessing the need for the training.

32. Which of the following is a reason for success of a training program?

a. Inadequacy of the leader/trainer

b. Different knowledge levels of the participants

c. Training is problem-oriented rather than technique-oriented

d. Complex language

33. Pretests and posttests are useful tools in a training program for:
    a. assessing learning and retention.
    b. assessing the learner's reaction to training.
    c. identifying the weakest employees.
    d. determining the cost of the training.

SOLUTIONS

1. c; Throughout the product life cycle, understanding and identifying customer needs is the first and most important step in any quality-related activity. [II.A.1]
2. a; The four tiers are *policies, procedures, instructions,* and *records.* The first or top tier represents policies. The second tier represents procedures. The third tier represents work instructions. The fourth tier represents a record of results of implementing a quality system. [II.B.1]
3. d; Compliance, accuracy, and clarity; without these components, users will lose confidence in the documentation. [II.B.1]
4. b; Effective implementation of a quality system that has been reviewed and approved for release is one of the two major components of a quality program. The other component is documentation of procedures and instructions of the quality system. [II.A.1]
5. c; Top management is responsible for endorsing the content of the final document as well as authorizing its release for use. [II.B.2]
6. a; Revision control and ensuring obsolete documents are not available for use are important principles of document control. [II.B.2]
7. b; ISO 9001:2015 is an international standard. [II.C]
8. d; Quality management is one of the standards, while the others are subjects of other standards (environmental management, risk management) or a quality management principle (continuous improvement). [II.C]
9. a; Customer focus is a key quality management principle emphasized in the ISO 9000 family. Options b and c refer to criteria for all ISO 9000 terms and definitions—and are illustrations of what not to do. In writing terms and definitions, technical definitions should be avoided, and coherent vocabulary agreed on and understood by all users (current and potential). Option d is a poor quality management principle. [II.C]

10. c; ISO 9004 consists of suggestions only, guidelines but not requirements for certification. [II.C]

11. b; There are seven criteria: Leadership; Strategy; Customers; Measurement, Analysis, and Knowledge Management; Workforce; Operations; and Results. [II.C]

12. a; If the auditor is an employee of the company to be audited, then we refer to this as an *internal* audit. It is necessary that the person who will conduct this audit is not related to or involved in the activity that is being audited. [II.D.1]

13. a; A quality system audit is an audit of the entire system. The other three are more specific and smaller in scope. [II.D.1]

14. c; Internal audits can be used to determine improvement opportunities for the quality management system. The other options apply to various types of external audits. [II.D.1]

15. a; The organization or company being audited (auditee) is responsible for making sure the corrective measures identified are addressed. [II.D.2]

16. b; Actually taking corrective action when nonconformities have been identified is the one step in the process that is most often poorly executed. [II.D.3]

17. a; Generally, an audit is carried out in three steps: (1) pre-examination—the auditor explains the objectives of the audit, introduces team members to senior management, and so on; (2) a "suitability" audit is conducted—comparison of documented procedures to the reference standard selected; and (3) implementation of the quality system is examined in depth, which may involve recommending corrective measures. [II.D.3]

18. c; The lead auditor is responsible for the accuracy as well as the validity of the final report. [II.D.2]

19. b; According to *The ASQ Certified Quality Engineer Handbook*, Fifth Edition, when assessing audit nonconformities, it is recommended to use a risk-based approach. For example, delineating between a major and minor nonconformance should be based on the level of risk and the analysis of audit objective evidence. [II.D.3]

20. c; According to *The ASQ Certified Quality Engineer Handbook*, Fifth Edition, proper planning is a key factor in achieving an efficient quality audit. Planning should be conducted with consideration of the client expectations, which include the scope, depth, and time frame. The lead auditor has responsibility for planning and conducting the audit and should be authorized to perform these activities. [II.D.3]

21. a; A follow-up audit is necessary to determine whether the problems and recommended corrections have been addressed. All three parties should arrange and agree on a follow-up audit. [II.D.4]

22. c; An increase in prevention costs, resulting in a decrease in defects, nonconformities, and so on, will result in a decrease in failure costs due to the reduction of defects and nonconformities. [II.E]

23. c; Since the data error correction, reprocessing, and reeducation occur before delivery of the product or service, the cost incurred would be an *internal* failure cost. [II.E]

24. b; Costs associated with activities performed to uncover or identify deficiencies (such as in personnel proficiency) are considered *appraisal* costs. [II.E]

25. a; Costs associated with activities related to supplier reviews are known as *prevention* costs. [II.E]

26. b; As failures are uncovered and corrected, appraisal activities will generally decrease, although appraisal activities will not and should not be completely eliminated. The type of appraisal activity can change. [II.E]

27. b; Depreciation of test equipment is a cost associated with measuring products to ensure conformance with quality standards and performance requirements, so it is an appraisal cost. [II.E]

28. c; The items that would be considered for appraisal costs include test equipment and material inspection, so the cost would be $12,000 + $65,000 = $77,000. [II.E]

29. a; None of the costs listed are external failure costs. In-plant scrap and rework is an internal failure cost. [II.E]

30. a; Quality system audits fall under prevention costs. [II.E]

31. d; Before developing and implementing any training program, a needs analysis should be conducted to determine whether the training is actually needed and, if so, what it should consist of. [II.F]

32. c; Frank Gryna noted that it was best if the participants could apply the training during the course to a problem of interest (be problem oriented). [II.F]

33. a; Pretests and posttests can be very useful for determining participant learning and retention of material. [II.F]

Part III

Product, Process, and Service Design

(58 questions)

QUESTIONS

1. Which of the following quality characteristics is an example of attribute data?
   a. Body temperature
   b. Number of defects
   c. Length of a bolt
   d. Width of a board
2. A part will be inspected using a limit gage. An acceptable part is one that:
   a. mates with both ends of the limit gage (both "go" and "no-go").
   b. mates only with the "no-go" end of the gage.
   c. mates with neither end of the limit gage.
   d. mates with only the "go" end of the limit gage.
3. A system has two components in parallel, each with a reliability of 0.95. The system reliability is approximately:
   a. 0.0975.
   b. 0.9500.
   c. 0.9975.
   d. 0.9999.
4. Which concept is used to identify drivers of quality based on the Voice of the Customer?
   a. Failure Mode and Effects Analysis
   b. Critical to Quality Tree
   c. Design of Experiments
   d. Cost of Quality

5. Consider the product/process development life cycle. The phase in which the product or process is in the hands of the customer for use is the:
   a. concept phase.
   b. design and development phase.
   c. production phase.
   d. post-production phase.
6. Which of the following would be a Critical to Quality characteristic?
   a. Length 18.0 to 18.2 cm
   b. Low costs
   c. No scrap
   d. Operating without failures
7. Consider the product/process development life cycle. The phase in which all possible solutions are explored is referred to as the:
   a. specification phase.
   b. prototype phase.
   c. concept phase.
   d. distribution phase.
8. Which of the following is an example of the Voice of the Customer in a Critical to Quality Tree?
   a. Process cycle time in seconds
   b. An insurance policy
   c. Mistakes as in defects per million opportunities
   d. Number of dropped calls
9. A tool that is useful for converting customer needs into design concepts is:
   a. process validation.
   b. quality function deployment.
   c. statistical process control.
   d. acceptance sampling.

10. A planning technique that translates customer requirements into technical requirements is:

    a. benchmarking.

    b. SWOT.

    c. customer satisfaction surveys.

    d. quality function deployment.

11. "House of quality" is a key diagram in:

    a. quality function deployment.

    b. flowcharts.

    c. PERT.

    d. CPM.

12. Which of the following methodologies may include components such as design for cost, manufacturability, or reliability?

    a. Design for Total Quality Management

    b. Design for Kaizen

    c. Design for Plan-Do-Check-Act

    d. Design for Six Sigma

13. Design reviews should be conducted by

    a. top management only.

    b. a quality engineer.

    c. a cross-functional team including quality, manufacturing, and suppliers, and so on.

    d. customers only.

14. For each design review conducted, there are many different points of view from which the design should be considered. One of these viewpoints is from a reliability standpoint. Which of the following would be a question from a reliability viewpoint?

    a. Can the design be sufficiently tested?

    b. Is the failure rate acceptably low?

    c. Can the product be shipped without damage?

    d. Can the parts needed for the design be obtained at an acceptable cost?

15. A specification is given as $8^{+0.05/-0.05}$. The limits for this specification are:

    a. 7.95 and 8.00.

    b. 7.95 and 8.05.

    c. 8.00 and 8.05.

    d. 8.05.

16. Consider the following control frame.

∥	Ⓛ	0.01	A	B	C

    0.01 represents the:

    a. primary datum.

    b. geometric feature or characteristic.

    c. geometric tolerance.

    d. modifier.

17. Consider the following control frame.

∥	Ⓛ	0.01	A	B	C

    "A" represents the:

    a. primary datum.

    b. secondary datum.

    c. tertiary datum.

    d. modifier.

18. Which of the following involves demonstrating that all equipment being used in a process has been installed properly (as per manufacturer's specifications)?

    a. Installation qualification

    b. Operational qualification

    c. Performance qualification

    d. Software qualification

19. For a process, confirming that *specified* requirements have been fulfilled is known as:

    a. validation.

    b. verification.

    c. qualification.

    d. performance.

20. Validation of a product, process, or service is normally carried out:

    a. during production or development.

    b. when customer requirements are determined.

    c. after completion of the product, process, or service.

    d. prior to development.

21. When can the reliability of a product be measured?

    a. At project initiation

    b. During development

    c. At the time of its release

    d. After it is released

22. The cumulative distribution function of failure at time t for a type of air conditioner is defined as $F(t) = 4/5{,}000$. Which of the following statements is (are) true?

    a. The probability that a randomly selected air conditioner fails by time t is 4,996/5,000.

    b. The probability that a randomly selected air conditioner will instantly fail is 4,996/5,000.

    c. The fraction of all units in the lot of air conditioners that fail by time t is 4/5,000.

    d. The probability that a randomly selected air conditioner will instantly fail is 4/5,000.

23. The cumulative distribution function of failure at time t for a computer model is 0.992. What fraction of this computer model fails by time t?

    a. 0

    b. 0.008

c. 0.992

d. 1

24. Which of the following probability distributions is the only one to have a constant failure rate?

a. Uniform

b. Normal

c. Exponential

d. Weibull

Use the following information for questions 25–27.

450 electrical components for a machine were subjected to a reliability test. The observed failures during 100-hour intervals are shown in the following table.

Upper bound (hours)	Number of failures, x
0	0
100	7
200	16
300	19
400	27

25. What is the estimated reliability at time $t = 200$?

a. 0.0004

b. 0.0156

c. 0.0511

d. 0.9489

26. What is the estimated failure density for time $300 < t < 400$?

a. 0.0006

b. 0.0933

c. 0.8467

d. 0.9067

27. What is the instantaneous rate of failure for time $300 < t < 400$?

    a. 0.00066

    b. 0.09333

    c. 0.15333

    d. 0.90667

28. A series system is composed of five subsystems, each with reliability 0.995. What is the reliability of the system?

    a. 0.1990

    b. 0.9752

    c. 0.9950

    d. 0.9999

29. A parallel system is composed of five subsystems, each with reliability 0.995. The reliability of this system is closest to which of the following values?

    a. 0.9500

    b. 0.9752

    c. 0.9950

    d. 1.0000

30. In a series system consisting of three subsystems, each with a reliability of 0.75, if one subsystem fails, what is the probability that the entire system will fail?

    a. 0

    b. 0.25

    c. 0.75

    d. 1.00

31. What is the reliability of a series system with two subsystems, each with a reliability of 0.80?

    a. 0

    b. 0.36

    c. 0.64

    d. 0.80

32. A machine has six subsystems and will fail when three or more of these subsystems fail. Each subsystem has a reliability of 0.88. What is the reliability of the machine?

    a. 0.4644

    b. 0.5997

    c. 0.9975

    d. 0.9998

33. A machine has a standby system design with three identical components in standby mode. Each component has a constant rate of failure $\lambda = 0.08$. What is the reliability of the machine at 100 hours of continuous operation?

    a. 0.0001

    b. 0.0424

    c. 0.7164

    d. 0.9999

34. 750 machines were tested for five hours. 28 failures occurred. What is the failure rate for these machines?

    a. 0.0075

    b. 0.033

    c. 27

    d. 133

35. The reliability of a component at $t = 400$ hours is 0.645. Assuming that the reliability of this component is exponential, which of the following is closest to the mean time to failure (MTTF)?

    a. 0.0011

    b. 0.355

    c. 3

    d. 910

36. The MTTF for a model of dishwasher is 115 months. What is the reliability that a randomly selected dishwasher of this model type survives past 115 months? Assume that the reliability function is an exponential distribution.

    a. 0.009

    b. 0.368

c. 0.632

d. 2.718

37. The MTTF for a machine is 525 hours. What fraction of machines survives past 1,000 hours? Assume that the reliability of the machine has an exponential distribution.

a. 0.19%

b. 14.88%

c. 85.12%

d. 99.81%

38. What is the probability a randomly selected rechargeable battery fails by six hours if the mean time between failures (MTBF) is five hours? Assume that the reliability of the component is exponentially distributed.

a. 0.1667

b. 0.3012

c. 0.6988

d. 0.8333

39. If the MTBF of a component is 900 hours, what is the reliability of this component at 500 hours? Assume that the reliability of the component is exponentially distributed.

a. 0.1653

b. 0.4262

c. 0.5738

d. 0.8347

40. The failure rate of a machine is 0.0008. Assuming the life distribution of this type of machine is exponential, what fraction of machines in the population survives longer than the MTTF?

a. 36.79%

b. 63.21%

c. 99.92%

d. 1,250%

41. The failure rate of a machine is 0.0008. Assuming the life distribution of this type of machine is exponential and the MTTR is five hours, what is the steady-state availability of this machine?

    a. 0.004

    b. 0.996

    c. 1,250

    d. 1,255

42. In a test of a repairable component of a machine, 600 components were tested for 25 hours, and 515 components failed. If the mean time to repair (MTTR) these components is eight hours, what is the steady-state availability of this component?

    a. 0.0343

    b. 0.2155

    c. 0.7845

    d. 29.13

43. Which type of maintenance policy creates a situation where there will be the maximum run time between repairs?

    a. Preventive maintenance policy

    b. Corrective maintenance policy

    c. Predictive maintenance policy

    d. Risk-based maintenance policy

44. A preventive maintenance policy is recommended if:

    a. the cost to repair the system after its failure is less than the cost of maintaining the system before its failure.

    b. the failure rate of the function of the system is monotonically decreasing.

    c. the cost to repair the system after its failure is greater than the cost of maintaining the system before its failure.

    d. the failure rate of the function of the system is constant.

45. In the second region identified in the general failure rate model (the bathtub curve), what is the most common cause for failures?

    a. The manufacturing process

b. Fatigue

c. Random failures

d. Out-of-box failures

46. Which of the following probability distributions is often used to model the second region of the bathtub curve?

a. Uniform

b. Weibull

c. Exponential

d. Normal

47. Which of the following probability distributions is often used to model the third region of the bathtub curve?

a. Uniform

b. Weibull

c. Exponential

d. Normal

48. What is the main goal of FMEA?

a. To help reduce the negative effects of potential failures before they occur

b. To organize the workplace environment

c. To communicate hierarchical relationships between events

d. To reduce the number of defects produced in a process

49. A hospital is preparing to do an FMEA for a surgical procedure. Which of the following groups of employees should make up the FMEA team members?

a. A doctor that performs this surgical procedure

b. A doctor and a nurse

c. A doctor, nurse, and business office employee

d. A doctor, nurse, technician, engineer, business office employee, and manager

50. Which risk management tool contains the following elements: potential failure types, the risk associated with the failures, and risk mitigations?

a. Fault tree analysis (FTA)

b. Failure mode and effects analysis (FMEA)

c. Risk management plan (RMP)

d. Risk management report (RMR)

51. An FMEA team has identified a potential failure and assessed the following score for severity, occurrence, and detection for this failure mode: 8, 2, 7. What is the RPN for this failure mode?

a. 16

b. 17

c. 56

d. 112

52. In the RPN above, if the risk reduction required is 40, which of the following risk ratings would indicate a situation requiring risk reduction?

a. Severity = 5, Occurrence = 2, Detection = 3

b. Severity = 5, Occurrence = 3, Detection = 2

c. Severity = 4, Occurrence = 3, Detection = 3

d. Severity = 3, Occurrence = 4, Detection = 4

53. Which of the following risk analyses considers product failures by a customer due to human factors?

a. FMECA

b. uFMEA

c. pFMEA

d. PHA

54. Which of the following types of FMEA is created after the others?

a. Process

b. Use

c. System

d. Design

55. Which of the following types of failure mode and effects analyses would best be used to assess the risk of insufficient product instructions in detail?

a. Design

b. Process

c. System

d. Use

56. In addition to the two main components of risk, the FMEA technique may add a rating scale for:

    a. detection.

    b. probability of occurrence.

    c. impact.

    d. strategic alignment.

57. An event that would be analyzed for impact and likelihood in a use failure mode and effects analysis (uFMEA) is:

    a. insufficient material strength.

    b. manufacturing process errors.

    c. adequacy of visual alarms.

    d. electrical failures.

58. What is the main goal of an FMECA?

    a. To reduce the number of defects produced in a process

    b. To develop a prioritization scheme for corrective action plans

    c. To organize the workplace environment

    d. To communicate hierarchical relationships between events

SOLUTIONS

1. b; A *defect* is an attribute. Either the defect is present or it is not. A defect has no real measurement on a continuous scale. [III.A]

2. d; An acceptable part should only fit within the "go" end of the limit gage. If it passes on both ends, it is an unacceptable part. [III.A]

3. c; For components in parallel, each with reliability of p, the system reliability is given by $R_s(t) = 1 - (1 - p)^2 = 1 - (1 - 0.95)^2 = 0.9975$. [III.E.2]

4. b; A Critical to Quality Tree is used to identify the drivers of quality based on the Voice of the Customer. [III.B.1]

5. d; The post-production phase of the development life cycle includes release of the product/process to the customer for its intended use or function. Normal wear and maintenance occur in this phase as well. [III.B.1]

6. a; A Critical to Quality characteristic is needed to achieve the driver and may consist of a specification such as length 18.0 to 18.2 cm. [III.B.1]

7. c; In the concept phase, all possible solutions are explored. Solutions that are not feasible are eliminated. [III.B.1]

8. b; An insurance policy is what the customer asks for. Voice of the Customer may be vague and not measurable, unlike the Critical to Quality characteristics. [III.B.1]

9. b; QFD, which uses the "house of quality," is helpful in turning customer needs into product characteristics or design concepts. [III.B.2]

10. d; QFD is a powerful tool for incorporating the voice of the customer and identifying the necessary technical requirements that will be necessary to satisfy customer requirements. [III.B.2]

11. a; The house of quality is a diagram that incorporates the key relationships in the process. [III.B.2]

12. d; Design for Six Sigma is an approach used to eliminate problems before they occur. Design for X is a subset of design for Six Sigma. [III.B.2]

13. c; Because of the complexity of design reviews, a cross-functional team consisting of many different personnel should conduct the review. The team should include, but not be limited to, quality personnel, manufacturing, suppliers, customers, and so on. [III.B.3]

14. b; Reliability includes metrics such as failure rates for the design under consideration. [III.B.3]

15. b; This is a bilateral tolerance with allowable variation in both directions. [III.C]

16. c; 0.01 represents the geometric tolerance for the feature on the part. [III.C]

17. a; "A" represents the primary datum. [III.C]

18. a; Installation qualification. The goal is to make sure the equipment involved is installed properly. [III.D]

19. b; Verification deals with determining whether the product or process satisfies specified requirements, that is, it meets all specifications or requirements. But this does not guarantee that the product or process specifications meet or exceed predetermined requirements (that is, it may not work like it is supposed to). [III.D]

20. c; Validation takes place after the product, process, or service is finished or completed. Validation is used to determine whether predetermined requirements are being met (product/process/service is working as intended). [III.D]

21. d; Reliability cannot be measured at any time prior to the release of the product; it can only be predicted. This is because the reliability of a product is dependent on many factors, including the system design, the reliability of its components, the operating environment, environmental factors, and manufacturing defects. [III.E.1]

22. c; The cumulative distribution function (cdf) of failure $F(t)$ has two interpretations: (1) the probability that a randomly selected unit from the population will fail by time t, or (2) the fraction of all units in the population that fail by time t. The hazard rate function $h(t)$ defines the instantaneous failure rate at time t. [III.E.1]

23. c; The cumulative distribution function (cdf) of failure can be interpreted as the fraction of all units in the population that fail by time t. Since $F(t) = 0.992$ for this computer model, the fraction of this model that fail by time t is 0.992. [III.E.1]

24. c; The exponential distribution is the only probability distribution to have a constant failure rate. For an exponential distribution with parameter λ, the hazard rate function $h(t) = \lambda$. [III.E.1]

From the following table, we can find the number of survivors up to time t to find the solutions to questions 25–27.

Upper bound (hours), t_i	Number of failures, x_i	Number of survivors, n_i
0	0	450
100	7	443
200	16	427
300	19	408
400	27	381

25. d; The estimated reliability at time $t = 200$ is

$$\hat{R}(t) = \frac{n_2}{N} = \frac{427}{450} = 0.9489.$$

[III.E.1]

26. a; The failure density is the probability density function defined as

$$\hat{f}(t) = \frac{n_3 - n_4}{(t_4 - t_3) \times N}.$$

Therefore,

$$\hat{f}(t) = \frac{408 - 381}{(400 - 300) \times 450} = 0.0006.$$

[III.E.1]

27. a; The instantaneous rate of failure is the hazard rate function $h(t)$, where

$$\hat{h}(t) = \frac{\hat{f}(t)}{\hat{R}(t)} = \frac{n_3 - n_4}{(t_4 - t_3) \times n_3} = \frac{408 - 381}{(400 - 300) \times 408} = 0.00066.$$

[III.E.1]

28. b; A series system has n components connected end-to-end. If one component fails, the entire system fails. The reliability of a series system made up of n components is

$$R_s(t) = R_1(t) \times R_2(t) \times \ldots \times R_n(t).$$

For this series system, the system reliability, therefore, is

$$R_s(t) = 0995^5 = 09752.$$

[III.E.2]

29. d; A parallel system is made up of components so that if one or more paths fail, the remaining path(s) are still able to perform properly. Therefore, the system fails when all units fail. The reliability of a parallel system is

$$R_s(t) = 1 - [F_1(t) \times F_2(t) \times \ldots \times F_n(t)].$$

For this series system, the system reliability therefore, is

$$R_s(t) = 1 - [(1 - 0.995)]^5 = 0.999999.$$

The reliability of the system is essentially 1. [III.E.2]

30. d; A series system is made up of n components connected in series. Therefore, the system will fail if any of the components or subsystems fail. [III.E.2]

31. c; A series system is made up of n components connected end-to-end. The system reliability, therefore, is the product of the reliabilities of each subsystem. An implication is that the system reliability will be lower than the reliability of the subsystem with the lowest reliability. [III.E.2]

32. c; The machine has a *k*-out-of-*n* system design where $k = 3$ and $n = 6$. The system reliability of a *k*-out-of-*n* system is

$$R_s(t) = \Sigma_{i=k}^{n} \binom{n}{i} p^i (1-p)^{n-i}$$

where $p = 0.88$.

Therefore, the system reliability is

$$R_s(t) = \binom{6}{3}(0.88)^3(0.12)^3 + \binom{6}{4}(0.88)^4(0.12)^2 + \binom{6}{5}(0.88)^5(0.12)^1 + \binom{6}{6}(0.88)^6(0.12)^0$$

$$= 20(0.88)^3(0.12)^3 + 15(0.88)^4(0.12)^2 + 6(0.88)^5(0.12)^1 + 1(0.88)^6(0.12)^0$$

$$= 0.02355 + 0.1295 + 0.3799 + 0.4644$$

$$= 0.9975.$$

[III.E.2]

33. b; The reliability of a standby system with *n* standby components is

$$R_s(t) = e^{-\lambda t} \sum_{i=0}^{n} \frac{(\lambda t)^i}{i!}$$

In this case, the machine is a standby system where $n = 4$, $\lambda = 0.08$, and $t = 100$.

Therefore, the system reliability is

$$R_s(t) = e^{-(0.08)(100)} \sum_{i=0}^{4} \frac{(0.08 \times 100)^i}{i!} = e^{-8} \sum_{i=0}^{4} \frac{8^i}{i!}$$

$$= e^{-8}\left[\frac{(8)^0}{0!} + \frac{(8)^1}{1!} + \frac{(8)^2}{2!} + \frac{(8)^3}{3!}\right] = 0.0424.$$

[III.E.2]

34. a; The failure rate, λ, is the number of failures divided by the total test time. Therefore,

$$\lambda = \frac{28}{750 \times 5} = 0.0075$$

Note that the mean time to failure (MTTF) is

$$\frac{1}{\lambda} \approx 133,$$

assuming the life distribution is exponential. [III.E.2]

35. d; We are given that $R(400) = 0.645$. Recall that MTTF = 1/failure rate = $1/\lambda$. Since the reliability of the component is exponential, we have

$$R(t) = e^{-\lambda t} \rightarrow R(400) = e^{-\lambda(400)} = 0.645$$

$$\ln\left(e^{-\lambda(400)}\right) = \ln(0.645)$$

$$-400\lambda = \ln(0.645)$$

$$\lambda = \frac{\ln(0.645)}{-400}$$

$$\lambda = 0.0011.$$

Therefore,

$$\text{MTTF} = \frac{1}{\lambda} = \frac{1}{0.0011} = 909.09 \approx 910 \text{ hours.}$$

[III.E.2]

36. b;

$$\text{MTTF} = 115 - \frac{1}{\lambda} = \frac{1}{\text{Failure rate}}.$$

This means that the failure rate, λ, is $\lambda = 1/115$. Since the reliability of the dishwasher has an exponential distribution, the probability that a randomly selected dishwasher survives past t months is $R(t) = e^{-\lambda t}$. Therefore, the probability that a randomly selected dishwasher survives past 115 months is

$$R\ (115) = e^{-115/115} = e^{-1} = 0.3679.$$

[III.E.2]

37. b; Another interpretation of reliability is the fraction of all machines that survives past a time t. We can find this fraction using the reliability function $R(t) = e^{-\lambda t}$, since

$$\text{MTTF} = 525,\ \lambda = \frac{1}{\text{MTTF}} = \frac{1}{525},\ R(1000) = e^{-\left(\frac{1000}{525}\right)} = 0.1488.$$

Approximately 15% of the machines will survive past 1,000 hours. [III.E.2]

38. c; The rechargeable battery is a repairable system (via recharging). Since the mean time between failures (MTBF) is five hours, the failure rate $\lambda = 1/5 = 0.2$. The probability that a battery fails by time t describes the failure function $F(t) = 1 - R(t) = 1 - e^{-\lambda t}$.

    Therefore,

$$F(6) = 1 - e^{\frac{6}{5}}.$$

    [III.E.2]

39. c; The mean time between failures (MTBF) is 900 hours. Therefore, the failure rate $\lambda = 1/900$. Assuming that the reliability of the component is exponential, the reliability at time $t = 500$ is defined as

$$R\ (500) = e^{-\lambda(500)} = e^{-500/900} = 0.5738.$$

    [III.E.2]

40. a; The reliability can be interpreted as the fraction of all components that survive past some time t. Therefore, $R(t) = e^{-\lambda t}$ and $R(\text{MTTF}) = e^{-\lambda\,(\text{MTTF})}$. Since the failure rate for this component is $\lambda = 0.0008$,

$$\text{MTTF} = \frac{1}{\lambda} = \frac{1}{0.0008} = 1250$$

$$R\ (\text{MTTF}) = R\ (1250) = e^{-0.0008(1250)} = 0.3679.$$

    36.79% of all components survive past the mean time to failure of 1,250 hours. [III.E.2]

41. b; The failure rate for this machine is $\lambda = 0.0008$, which means that the mean time between failures (MTBF) is

$$\frac{1}{\lambda} = \frac{1}{0.0008} = 1250 \text{ hours.}$$

    Since MTTR = 5 hours, the steady-state availability of the machine is

$$A = \frac{\text{MTBF}}{\text{MTBF} + \text{MTTR}} = \frac{1250}{1250 + 5} = 0.9960.$$

    [III.E.2]

42. c;

$$\lambda = \text{Failure rate} = \frac{\#\text{ failures}}{\text{Total test time}} = \frac{515}{600 \times 25} = 0.0343.$$

Therefore,

$$\text{MTBF} = \frac{1}{\lambda} = \frac{1}{0.0343} = 29.1262 \approx 29.$$

Since the mean time to repair the components is eight hours, the steady-state availability of the component is

$$A = \frac{\text{MTBF}}{\text{MTBF} + \text{MTTR}} = \frac{29.1262}{29.1262 + 8} = 0.7845.$$

[III.E.2]

43. b; A corrective maintenance policy does not require any repairs or other preventive maintenance until a failure occurs. Therefore, this type of policy allows for the maximum run time between repairs. [III.E.2]

44. c; A preventive maintenance policy is recommended when the cost to repair the system after its failure is greater than the cost of maintaining the system before its failure and the failure rate is monotonically increasing. If the failure rate is decreasing or constant, then the system is more likely to improve over time, and preventive actions would be a waste of resources. [III.E.2]

45. c; The second region of a bathtub curve is the constant failure rate region. This region is characterized by the inherent failure rate of the components of a product. [III.E.3]

46. c; The second region of the bathtub curve is characterized by a constant failure rate. The exponential distribution is the only probability distribution that has a constant failure rate; therefore, the exponential distribution is often used to model the second region of the bathtub curve. [III.E.3]

47. b; The third region of the bathtub curve, the wear-out region, is characterized by an increasing failure rate over time. This time-dependent failure rate is often modeled using the Weibull distribution. [III.E.3]

48. a; The main goal of an FMEA is to help a team identify and eliminate the negative effects of a potential failure before it occurs. [III.E.4]

49. d; An FMEA team should be composed of a cross section of employees across several fields. At a minimum, the FMEA team should have a doctor, nurse, technician, engineer, and manager. An FMEA team should have members that are both technical and nontechnical to have many viewpoints represented during the process. [III.E.4]

50. b; An FMEA contains all the components listed. The other tools are part of the risk management process but serve different purposes. [III.E.4]

51. d; The risk priority number (RPN) is determined by multiplying the three components of risk together: severity, occurrence, and detection. Therefore, RPN = 8 x 2 x 7 = 112. [III.E.4]

52. d; The first three options have RPNs less than 40, the threshold for required risk reduction. For option (d), RPN = 3 x 4 x 4 = 48, so risk reduction is required. [III.E.4]

53. b; A use FMEA (uFMEA) specifically considers customers' human factors issues. [III.E.4]

54. a; The process FMEA (pFMEA) is initiated last, closer to the end of design and development, after the use FMEA (uFMEA), design FMEA (dFMEA), and system FMEA have already been created. The product must be designed before the process to make it. [III.E.4]

55. d; Product instructions are part of the user interface and are best analyzed in detail through a uFMEA.

56. a; The main components of risk are probability of occurrence (or likelihood) and impact (or severity). Detection is the third rating scale often used with FMEAs.

57. c; Visual alarms are part of the user interface that is evaluated via a uFMEA. The other options would be best considered via a dFMEA (options a and d) or pFMEA (option b).

58. b; An FMECA is an assessment of risk that provides a prioritization of corrective action based on severity and occurrence of failure modes. [III.E.4]

Part IV
Product and Process Control

(63 questions)

QUESTIONS

1. An employee works in an area of a facility where tools are painted. Which of the following would this employee find in the control plan for the process of painting tools?
   a. The sample size and frequency of sampling to monitor the painting process
   b. Employee training records
   c. Management review minutes
   d. Manufacturing costs
2. Which of the following would you typically find in a reaction plan for a process?
   a. Sampling plan
   b. Control chart
   c. Detailed methods to contain questionable products
   d. Gage R&R study
3. Why are work instructions an important component of process control?
   a. They explain sampling procedures.
   b. They identify bottlenecks in a process.
   c. They help reduce process errors.
   d. They help to manage quality costs.
4. Why is it important to identify materials used in producing a product?
   a. To help manufacturers reduce risk
   b. To save money

c. To help ensure product reproducibility

d. To ensure management support

5. Which of the following is one of the most effective methods of accounting for and tracking product materials?

a. Control charts

b. OC curves

c. Paper move tags

d. Radio frequency identification

6. A food company is planning to release a new type of cookie. At what stage in the process should the company consider tracing the cookie's components?

a. Before the release of the cookie to the public

b. After any inquiries from customers

c. At the beginning of the cookie's development

d. Only when there is an increase in nonconforming cookies

7. Keeping nonconforming products from being mixed with good/conforming products within a manufacturing plant is known as:

a. compliance.

b. material segregation.

c. material review.

d. defect classification.

8. A recall due to a brake malfunction on an automobile was issued by a car company. How would the company classify this type of defect?

a. Minor defect

b. Major defect

c. Serious defect

d. Critical defect

9. Customers have complained about the battery life for a new line of computers released. How would the computer company classify this type of product defect?

a. Minor defect

b. Major defect

c. Serious defect

d. Critical defect

10. What purpose does the material review board of a company serve?

    a. Oversee the traceability of a product

    b. Develop sampling plans for product materials

    c. Identify the source of all product components

    d. Determine corrective action for nonconforming components

11. Acceptance sampling is most appropriate for inspecting products when:

    a. the product is complex.

    b. the fraction of nonconforming products is unknown.

    c. there are high inspection costs.

    d. the cost of making a wrong decision is high.

12. A disadvantage of sampling inspection compared to 100% inspection is:

    a. sampling has lower inspection costs.

    b. sampling is more time-intensive than 100% inspection.

    c. sampling emphasizes that the supplier is concerned with quality.

    d. rejecting a lot that conforms to specification as a result of variability in sampling.

13. Which of the following tools is most appropriate when evaluating a sampling plan?

    a. Scatter plot

    b. Operating characteristic (OC) curve

    c. Process map

    d. Affinity diagram

14. A company requires a part on a small lot-by-lot basis. On which of the following probability distributions should the OC curve of the sampling plan be based?

    a. Hypergeometric distribution

    b. Binomial distribution

c. Poisson distribution

d. Normal distribution

15. A company is deciding between two competing sampling plans for a product they receive from a manufacturer. The diagram below displays the OC curves for these two sampling plans.

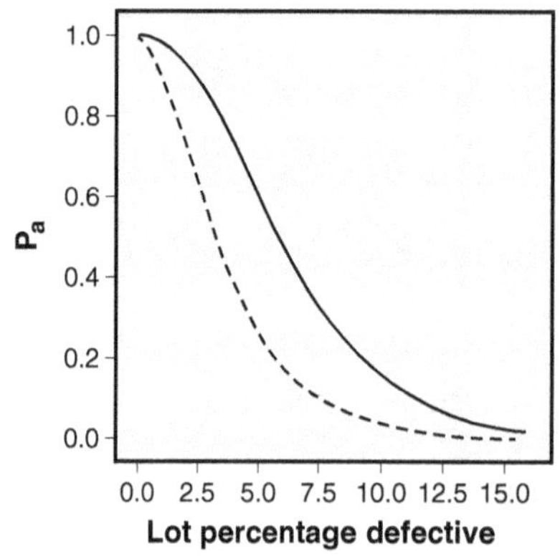

Compared to the solid line, the dashed OC curve:

a. is better for the company.

b. will accept lots with a higher probability.

c. is better for the manufacturer.

d. provides equal risk for the company and manufacturer.

16. Consider a sampling plan with $n = 45$ and $c = 2$. Given that the percentage nonconforming is 0.05 and a binomial distribution is appropriate, what is the probability of accepting the lot?

a. 0.3350

b. 0.3923

c. 0.6077

d. 0.6093

17. In acceptance sampling, what does the acronym "LTPD" mean?

a. The quality level

b. The poorest quality in an individual lot that should be accepted

c. The average quality of outgoing products

d. The worst tolerable process average of a continuing series of lots

18. Type I error in hypothesis testing corresponds to which of the following in acceptance sampling?

    a. Consumer's risk

    b. Producer's risk

    c. Beta risk

    d. High risk

19. Consider a sampling plan with $n = 45$ and $c = 2$. Given that the percentage nonconforming is 0.05 and a lot size is assumed to be infinite, what is the average outgoing quality?

    a. 0.0304

    b. 0.3923

    c. 0.6077

    d. 0.9696

20. In an acceptance sampling plan, how does the probability of acceptance change as the sample size increases, assuming the acceptance number and lot size are held constant?

    a. P_a increases.

    b. P_a decreases.

    c. P_a stays constant.

    d. Not enough information is provided.

21. Holding all else equal, which factor has the least effect on the OC curve?

    a. The lot size, N

    b. The sample size, n

    c. The acceptance number, c

    d. The probability of acceptance, P_a

22. What is a potential disadvantage of single sampling plans?

    a. They do not use tightened inspection.

    b. They are hard to administer in practice.

    c. The sample size is adjustable for changes in incoming product quality.

    d. They are more complicated to use.

23. The attribute sampling plan for a specific product utilizes an acceptance value of 2 and rejection value of 5. The inspector discovers two defective items in the current sample. What is the next appropriate action?

    a. Accept the lot, since the number of defective items does not exceed the Ac value.

    b. Accept the lot, since the number of defective items is less than the Re value.

    c. Reject the lot, since the number of defective items is not less than the Ac value.

    d. Draw another sample.

24. A double sampling plan is implemented with the following parameters:

$$n_1 = 50,\ c_1 = 1,\ r_1 = 4,\ n_2 = 75,\ c_2 = 3,\ r_2 = 6$$

    What is the probability that a decision is made (to accept or reject the lot) on the first sample if the percentage nonconforming, p, is equal to 0.10?

    a. 0.2165

    b. 0.7835

    c. 0.8614

    d. 0.9221

25. A double sampling plan is implemented with the following parameters:

$$n_1 = 50,\ c_1 = 1,\ r_1 = 3,\ n_2 = 75,\ c_2 = 3,\ r_2 = 4$$

    What is the average sample number for this sampling plan if the percentage nonconforming, p, is equal to 0.10?

    a. 50

    b. 67

    c. 109

    d. 125

26. When compared with sampling plans with equal protection, the average sample number of double sampling plans generally is:

    a. smaller when the quality is poor.

    b. larger when the quality is poor.

c. larger when the quality is near the indifference level ($p = 0.50$).

d. smaller when the quality is near the indifference level ($p = 0.50$).

27. Which of the following types of sampling are allowed in ANSI/ASQ Z1.4-2003 (R2018)?

    a. Single, double, and sequential

    b. Single, double, and multiple

    c. Single, sequential, and stratified

    d. Double, sequential, and multiple

28. A lot size of 1,500 is to be inspected using ANSI/ASQ Z1.4-2003 (R2018). The AQL is 0.40%. What is the required sample size for a single sampling plan (normal, level II inspection)?

    a. 80

    b. 125

    c. 200

    d. 315

29. A lot size of 500 is to be inspected using ANSI/ASQ Z1.4-2003 (R2018). The AQL is 0.65%. What are the required sample size, acceptance number, and rejection number for a single sampling plan (normal level II inspection)?

    a. $n = 50$, Ac = 0, Re = 1

    b. $n = 50$, Ac = 1, Re = 2

    c. $n = 80$, Ac = 0, Re = 1

    d. $n = 80$, Ac = 1, Re = 2

30. A lot of size 200 is to be inspected using ANSI/ASQ Z1.4-2003 (R2018). The AQL is 0.65%. What are the required sample sizes for a double sampling plan (normal, level II inspection)?

    a. 13, 13

    b. 20, 20

    c. 20, 40

    d. 32, 32

31. A lot of size 400 is to be inspected using ANSI/ASQ Z1.4-2003 (R2018). The AQL is 0.65%. For the first sample of a double sampling plan (normal, level II inspection), what are the acceptance and rejection numbers?

    a. Ac = 0, Re = 2

    b. Ac = 0, Re = 3

    c. Ac = 1, Re = 2

    d. Ac = 1, Re = 4

32. An advantage of variables sampling plans compared to attribute sampling plans is that variables sampling plans:

    a. require smaller sample sizes with equal protection compared to attribute sampling plans.

    b. are easier to implement.

    c. can monitor multiple quality characteristics at once.

    d. are less expensive to implement.

33. A lot of 100 items is submitted for inspection under ANSI/ASQ Z1.9-2003 (R2018), level II, normal inspection with AQL = 0.65%. Both upper and lower specification limits are specified for these items. What sample size is required for this sampling plan?

    a. 5

    b. 7

    c. 10

    d. 20

34. A lot of twenty items is submitted for inspection under ANSI/ASQ Z1.9-2003 (R2018), level II, normal inspection with AQL = 1.00%. The required sample size for this sampling plan is four. The results of the four measurements are: 51, 47, 45, 39. Does this lot meet the acceptance criteria? The upper and lower specification limits for this process are 52 and 37, respectively.

    a. Yes, since the total percentage nonconforming is less than the maximum allowable.

    b. No, since the total percentage nonconforming is less than the maximum allowable.

c. Yes, since the total percentage nonconforming is more than the maximum allowable.

d. No, since the total percentage nonconforming is more than the maximum allowable.

35. Why is sampling integrity important?

    a. To be able to accurately read an OC curve

    b. To ensure that a product has not been contaminated during inspection

    c. To have the most efficient process

    d. To reduce the amount of sampling required

36. Which of the following types of control are methods used to maintain sample integrity?

    a. Batch, change, and inspection

    b. Batch, configuration, and inspection

    c. Batch, change, and configuration

    d. Change, configuration, and inspection

37. Which of the following is a result of dimensional inspection of a product?

    a. Improved quality of the product

    b. Conformity to customer aesthetic requirements

    c. Determination of a conforming or nonconforming product

    d. Determination of employees' capabilities

38. Which of the following provides the most precise angular measurement?

    a. Protractor

    b. Bevel protractor

    c. Square

    d. Sine bar

39. The outside dimensions of a part are inspected for conformance to product specifications. Which of the following measuring tools is most appropriate for this measurement?

    a. Steel ruler

    b. Plug gage

c. Snap gage

d. Sine bar

40. A part was inspected using a go/no-go gage. The part did not mate with either end of the gage. Did the part pass the inspection?

a. No, the part should be rejected.

b. Yes, the part should be accepted.

c. Yes or no, depending upon further inspection.

d. Not enough information is provided.

41. Which of the following methods is the most common for surface measurements?

a. Reflectance meters

b. Optical interference

c. Pneumatics

d. Stylus averaging

42. A key component of a part is round in shape. Which of the following measurement tools is most appropriate to measure this component?

a. Ring gage

b. Vernier calipers

c. Precision spindle

d. Dial indicator

43. The maximum tensile strength a particular part can withstand is an important quality characteristic of this part. Which of the following is most appropriate to test this quality characteristic?

a. 100% inspection

b. Screening inspection

c. Coordinate measuring machine

d. Nondestructive testing

44. Which of the following nondestructive tests can only be used on parts made of iron and steel?

a. Liquid penetration testing

b. Magnetic particle testing

c. Eddy current testing

d. Ultrasonic testing

45. Which of the following nondestructive tests can only be used on parts made of conductive materials?

a. Liquid penetration testing

b. Magnetic particle testing

c. Eddy current testing

d. Radiographic testing

46. Which of the following organizations maintains the standards of measurement in the United States?

a. ISO

b. ASQ

c. NIST

d. SI

47. Physical artifacts, paper standards, and definition of base units are elements of what field?

a. Quality management

b. Auditing

c. Statistical sampling

d. Metrology

48. How does NIST define the kilogram, the base unit of mass?

a. As a physical artifact used for direct reference

b. The amount of water required to fill a 36-ounce beaker

c. Based on the weight of carbon molecules

d. A theoretical concept

49. Slight changes in temperature where parts are measured for conformance could lead to:

a. systematic error in measurement.

b. random error in measurement.

c. operator error in measurement.

d. type II error in measurement.

50. To tie a laboratory's measurement standard to national standards, a company could use which of the following?

a. Measurement assurance

b. Calibration

c. Traceability

d. Gage R&R study

51. Information on the contribution to error of technicians or a laboratory's procedures can typically be found in:

a. measurement calibration records.

b. measurement traceability protocols.

c. the material review board.

d. measurement assurance protocols.

52. The goal of equipment calibration is to maintain the system's:

a. accuracy.

b. precision.

c. sensitivity.

d. readability.

53. While analyzing repeated measurements of a quality characteristic, an employee notices that the values are consistently larger than the reference value. Which of the following has occurred in this measurement process?

a. Random error

b. Systematic error

c. Type I error

d. Type II error

54. Which of the following components of a measurement system contributes to its accuracy?

a. Repeatability

b. Reproducibility

c. Linearity

d. Random error

55. Which of the following increases measurement system error?

a. Bias

b. Correlation

c. Precision

d. Calibration

56. The length of a part should be 6.00 cm. A particular part is repeatedly measured six times, yielding the following measurements: 6.05, 5.94, 5.98, 6.01, 6.03, 5.97. What is the resulting bias for this part?

a. -0.003

b. 0.003

c. 0.041

d. 5.997

57. The lengths of three parts have the following reference values: 3.00 cm, 6.00 cm, 12.00 cm. Ten repeated measurements were taken for each type of part. The observed averages of these three samples were: 2.998, 5.843, 12.275. Which of the following statements is true?

a. There is evidence of nonlinearity in this measurement system.

b. There is evidence of stability issues in this measurement system.

c. The measurement system is not reproducible.

d. The measurement system is not repeatable.

58. The precision of a measurement system can be defined as:

a. the difference between the recorded measurement and the target value of a quality characteristic.

b. the variability present when the same part is measured multiple times under the same conditions.

c. the ability to get the same degree of bias in a measurement over time.

d. the ability to get the same degree of bias in a measurement for parts of different sizes.

59. In a gage R&R study, the variance associated with repeatability was found to be 0.8842 and the variance associated with reproducibility was found to be 0.2466. What is the variance associated with measurement error?

   a. 0.6376

   b. 0.7210

   c. 1.1308

   d. 1.4369

60. The ANOVA table below displays the results of a gage R&R study where two operators measured five parts three times each using the same gage.

Source	df	SS	MS	*F*	*p*
Part	4	44	11	12.69	0.015
Operator	1	0.033	0.033	0.04	0.854
Part × Operator	4	3.467	0.867	0.70	0.599
Error	20	24.667	1.233		
Total	29	72.167			

   Using this full model, what is the measurement error variability?

   a. 0

   b. 0.867

   c. 1.233

   d. 1.828

61. Based on the results from the gage R&R study in the previous question, which of the following statements is true?

   a. There is a significant interaction between parts and operators.

   b. More operator training is necessary in this process.

   c. This is a capable process.

   d. The variation associated with repeatability should be investigated.

62. Refer to the previous gage R&R study. Suppose that the upper and lower specification limits for the quality characteristic are 30 and 10, respectively. What is the precision-to-tolerance ratio for this process?

    a. 0.333

    b. 0.370

    c. 0.406

    d. 0.523

63. A gage R&R study was performed for a process using ten parts, two operators, and three replicates. The *R*-chart below displays information about the gage capability. What conclusions can be drawn from this control chart?

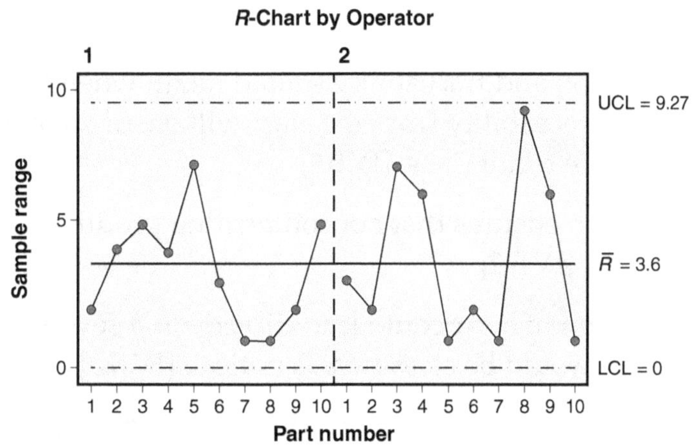

    a. There are significant differences between operators.

    b. The operators are consistent in their use of the gage.

    c. The gage is not able to distinguish between parts.

    d. The gage is able to distinguish between parts.

SOLUTIONS

1. a; Control plans typically include the following: a description of the process, any equipment used in the process, reference drawing numbers, the process characteristic that is being controlled, how the process characteristic should be evaluated, the sampling plan, the method of control, and a reaction plan if the process goes out of control. [IV.A]

2. c; A reaction plan typically contains four elements: containment, diagnosis, verification, and disposition. The reaction plan should therefore include specific directions for the operators on how to contain products that are suspected of being nonconforming or out of spec. [IV.A]

3. c; Work instructions provide details of a process for those who operate the process. Providing detailed instructions helps eliminate errors by providing all required machine settings, setup instructions, and visual aids. [IV.A]

4. a; Knowing where materials come from also gives the manufacturers the ability to trace the product. This helps to reduce exposure to additional risks. [IV.B.1]

5. d; Radio-frequency identification (RFID) is a common method used to identify and track materials using barcodes, scanners, and RFID receivers. RFID tags allow manufacturers to maintain detailed information about materials as well. [IV.B.1]

6. c; Product identification and traceability should begin when the process is being designed. Including traceability from the start will ensure adequate quality of the cookie and help reduce recall risks. [IV.B.1]

7. b; Material segregation ensures that nonconforming products are not mixed with conforming products. [IV.B.2]

8. d; A brake malfunction in a car could lead directly to a severe customer injury; therefore, this defect would be classified as critical. [IV.B.3]

9. b; The poor battery life of the computer reduces the usability of the product, but will not likely lead to an injury or significant economic loss. Therefore, the company would classify this as a major defect. [IV.B.3]

10. d; The material review board (MRB) determines the appropriate corrective action to take in the case that nonconforming components are discovered. It also determines any additional actions to take to prevent similar nonconformities in the future. [IV.B.4]

11. c; Acceptance sampling is most appropriate when the inspection costs are high or when destructive inspection is the only method of inspection. Acceptance sampling is also appropriate when the inspection of a product is monotonous and causes inspector fatigue, which could lead to an increase in inspection errors. [IV.C.1]

12. d; Sampling inspection has several advantages as it saves time and energy compared to 100% inspection. However, there is variability inherent in sampling. Therefore, there are chances of making type I and type II errors: rejecting lots that conform to the quality specifications or accepting lots that do not conform to quality specifications. [IV.C.1]

13. b; An operating characteristic (OC) curve displays the probability of accepting lots at varying levels of percentage nonconforming. It provides a simple, graphical way to evaluate a sampling plan. Scatter plots are appropriate to analyze the relationship between two variables. A process map is used to help understand the steps of a process. Affinity diagrams are useful as a planning tool to help organize ideas generated by a group. [IV.C.1]

14. a; Because the products are received on a lot-by-lot basis and are not part of a continuous process, the OC curve should be based on the hypergeometric distribution. This OC curve is called a type A OC curve. Type B OC curves are used to evaluate a sampling plan for a continuous process or for a process that has a large lot size. Type B OC curves are developed using the binomial or Poisson distributions. [IV.C.1]

15. a; The dashed OC curve is better for the company. As the percentage nonconforming increases, the probability of accepting the lot decreases much faster compared to the other OC curve. This means that the company will accept fewer unacceptable lots from the producer. [IV.C.1]

16. c; The probability of acceptance is defined as

$$P_a = P(d \le c) = \sum_{d=0}^{c} \binom{n}{d} p^d (1-p)^{n-d}.$$

Note that

$$\binom{n}{d} = \frac{n!}{d!(n-d)!}$$

In this example, $n = 45$, $c = 2$, and $p = 0.05$. So,

$$P_a = \sum_{d=0}^{2} \binom{45}{d} (0.05)^d (0.95)^{n-d} = 0.6077.$$

17. b; LTPD is the *lot tolerance percent defective,* expressed in percentage defective. It is the worst quality level in an individual lot that should be accepted when submitted for acceptance sampling. Consequently, the LTPD has a low probability of acceptance. [IV.C.1]

18. b; A type I error in hypothesis testing is the probability of rejecting the null hypothesis when in fact the null hypothesis is true. Therefore, type I error corresponds to the producer's risk in acceptance sampling: the probability that a lot is rejected when in fact the lot is within the acceptable quality level. [IV.C.1]

19. a; Assuming an infinite lot size, the *average outgoing quality* (AOQ) can be calculated as AOQ = $P_a p$. In this sampling plan,

$$P_a = \sum_{d=0}^{2} \binom{45}{d} (0.05)^d (0.95)^{n-d} = 0.6077.$$

Therefore, AOQ = (0.6077)(0.05) = 0.0304. [IV.C.1]

20. b; As the sample size increases, the probability of acceptance decreases. Consider, for example, an infinite lot size, an acceptance number $c = 1$, and fraction nonconforming $p = 0.05$. For various sample sizes n, the probability of acceptance is:

Sample size (n)	P_a
2	0.9975
5	0.9774
10	0.9139
20	0.7358

[IV.C.1]

21. a; The lot size, N, has the smallest effect on the OC curve compared to the acceptance number c, the sample size n, and the probability of acceptance P_a. [IV.C.1]

22. a; The sample size of a single sampling plan does not change, regardless of changes in incoming product quality levels. Therefore, reduced or tightened inspection is not implemented with single sampling plans, which typically reduces sampling costs. [IV.C.2]

23. a; When using a double sampling plan, after you take the first sample you will either accept the lot, reject the lot, or draw another sample. If the number of defective items in the first sample, d_1, is less than or equal to the acceptance number, then you accept the lot. If d_1 is greater than the rejection number value, then you reject the lot. If d_1 is more than the acceptance number but less than the rejection number, then you draw another sample. In this case, $d_1 = 2 = \text{Ac}$. Therefore, the lot is accepted, and another sample does not need to be drawn. [IV.C.2]

24. b; Let D_1 denote the probability that the lot is accepted or rejected on the first sample. Let d_1 denote the number of defects found in the first sample. The lot is accepted if there are one or fewer defects noted in the first sample. The lot will be rejected if there are four or more defects in the first sample. (Another sample will be drawn if two or three defects are found in the first sample.) Therefore,

$$P(D_1) = P(d_1 \le 1) + P(d_1 \ge 4) = 1 - P(d_1 = 2) - P(d_1 = 3)$$
$$= 1 - \binom{50}{2}(0.10)^2(0.90)^{48} - \binom{50}{3}(0.10)^3(0.90)^{47}$$
$$= 1 - 0.0779 - 0.1386 = 0.7835.$$

[IV.C.2]

25. b; Let P_2 denote the probability of requiring a second sample. From the previous problem (question 24), the probability of only drawing one sample is 0.7835. Therefore, the probability of requiring a second sample $P_2 = 1 - 0.7835 = 0.2165$. For a double sampling plan, the average sample number is defined as ASN = $n_1 + n_2P_2 = 50 + 75 \times 0.2165 = 66.24$ or 67. [IV.C.2]

26. a; When the quality of the lots is very good or very poor, the average sample size will generally be smaller compared with other sampling plans with equal protection. This is because the probability of making a decision on the first sample will tend to be higher, and thus fewer second samples are required. [IV.C.2]

27. b; ANSI/ASQ Z1.4-2003 (R2018) allows for single sampling, double sampling, and multiple sampling. [IV.C.2]

28. d; Given that the lot size is 1500, the sample size code letter from the ANSI/ASQ Z1.4-2003 (R2018) sampling tables (ANSI/ASQ Z1.4-2003 [R2018] Table 1) for normal, level II sampling is K. Following the down arrow and the guidance in ANSI Z1.4, the required sample size is $n = 315$ (from ANSI/ASQ Z1.4-2003 [R2018] Table II-A). [IV.C.2]

29. d; Given that the lot size is 500, the sample size code letter from the ANSI/ ASQ Z1.4-2003 (R2018) sampling tables for normal, level II sampling is H. Using ANSI/ASQ Z1.4-2003 (R2018) Table II-A, using the AQL of 0.65%, the downward arrow points to the sampling plan for code letter J. Therefore, the required sample size is 80, the acceptance number is 1, and the rejection number is 2. [IV.C.2]

30. b; Given that the lot size is 200, the sample size code letter from the ANSI/ASQ Z1.4-2003 (R2018) sampling tables for normal, level II sampling is G (Table 1). Using ANSI/ASQ Z1.4-2003 (R2018) Table III-A, Double Sampling Plans for Normal Inspection, and the same size code letter G, the sample size required is 20 for both the first and second sample regardless of the AQL. [IV.C.2]

31. a; Given that the lot size is 400, the sample size code letter from the ANSI/ASQ Z1.4-2003 (R2018) sampling tables for normal, level II sampling is H (Table 1). Using ANSI/ASQ Z1.4-2003 (R2018) Table III-A, Double Sampling Plans for Normal Inspection, the same size code letter H, and AQL = 0.65%, the downward arrow points to the sampling plan for code letter J. Therefore, the acceptance number for the first sample is 0 and the rejection number for the first sample is 2. [IV.C.2]

32. a; Variables control charts offer equal protection compared to an attribute sampling plan with a much smaller sample size. However, they are generally more expensive to implement and are much harder to administer. They also can only monitor one quality characteristic at a time. [IV.C.2]

33. c; ANSI/ASQ Z1.9-2003 (R2018) Table A-2 gives sample size code letter F. From ANSI/ASQ Z1.9-2003 (R2018) Table B-3, this code letter corresponds to a sample size of 10, regardless of the AQL level. [IV.C.2]

34. d; Note that USL = 52, LSL = 37. From the provided sample data, $\bar{x} = 45.5$, $s = 5.00$. Using this information, the two quality indices for the sampling plan are

$$Q_u = \frac{\text{USL} - \bar{x}}{s} = \frac{52 - 45.5}{5} = 1.30$$

and

$$Q_L = \frac{\bar{x} - \text{LSL}}{s} = \frac{45.5 - 37}{5} = 1.70.$$

From ANSI/ASQ Z1.9-2003 (R2018) Table A-2, the sample size code letter for this situation is C. Using Table B-3 (since there are double specification limits) and the AQL level of 1.00%, the maximum allowable percentage nonconformity (M) is $M = 1.49\%$. Table B-5 is then used to estimate the lot percentage nonconforming. The percentage above the specification limit with $n = 4$ and $Q_U = 1.30$ is 6.67%. The percentage below the specification limit with $n = 4$ and $Q_L = 1.70$ is 0.00%. The total percentage nonconforming is 6.67% + 0.00% = 6.67%. Since the percentage nonconforming is greater than M, the maximum allowable percentage nonconforming, the conclusion is that the lot is not acceptable. [IV.C.2]

35. b; Sample integrity is necessary to ensure that a product has not been misused or contaminated and to ensure that sampling is accurate to the entire lot of products. [IV.C.3]

36. c; To maintain sample integrity, you can use batch control (which includes documenting the steps used to create the product, giving unique identification numbers to each batch produced, and maintaining a log listing the amount of each material inserted into the batch), change control (which involves classifying and tracking changes made to a product), and configuration control (an extension of change control). [IV.C.3]

37. c; Dimensional inspection involves measurement of a product to determine whether the product meets required specifications. Inspecting a product to determine if it meets the specifications, however, does not determine its quality. The quality of the product will depend on the engineering design and how it was manufactured. [IV.D.1]

38. d; A sine bar is more precise for measuring angles as compared to protractors, bevel protractors, or squares. [IV.D.1]

39. c; Gages are used to determine conformance (or nonconformance) of a part's dimensions to specifications. Specifically, snap or ring gages are used to check outside dimensions; plug gages are used to check inside dimensions. [IV.D.1]

40. a; A part passes inspection with a go/no-go gage if the part mates with the go end and does not mate with the no-go end. Since this part did not mate with either end, it did not pass the "go" inspection. Therefore, the part should be rejected. [IV.D.1]

41. d; Using a stylus to measure an average electrical signal is the most common method of surface measurement because it is fast, repeatable, interpretable, and relatively inexpensive. [IV.D.1]

42. c; Precision spindle instruments are used for roundness measurements. Calipers are used for linear measurements, ring gages are used to assess specification conformance of outside dimensions, and dial comparators are used to check operations (for example, verify alignments). [IV.D.1]

43. d; Measuring the maximum tensile strength of a part would be destructive to the part. Therefore, screening inspection or 100% inspection is not appropriate; nondestructive testing is required. [IV.D.2]

44. b; Magnetic particle testing is used on parts that can be magnetized; therefore, magnetic particle testing can be done only on parts made of iron, steel, or allied materials. [IV.D.2]

45. c; Eddy current testing applies an AC current through a coil near the surface of the part under inspection; therefore, only parts made of conducting materials can be tested with eddy current testing. [IV.D.2]

46. c; The National Institute of Standards and Technology (NIST) maintains the standards of measurement in the United States. [IV.E]

47. d; All these areas must be taken into account in metrology. Physical artifacts must be manufactured with high precision (for example, gage blocks) to be used as masters for reference. Paper standards include the published documents that contain the specifications or accepted methods for taking measurements. Definition of base units is critical as they form a reference from which all other units are derived. [IV.E]

48. a; The kilogram is the only base unit that is defined in terms of a physical artifact, which is made of a platinum-iridium alloy and protected by the International Bureau of Weights and Measures in Paris. A duplicate is also protected by NIST in the United States. [IV.E]

49. b; Errors in measurement are classified as *random* errors (accidental fluctuations that cannot be predicted) or *systematic* errors (for example, operator error or calibration error). Ambient changes like slight temperature changes in a laboratory are classified as random error. [IV.E]

50. c; Traceability is intended to tie a laboratory's measurement standards to national standards. [IV.E]

51. d; Measurement assurance protocols provide details on the accuracy of a measuring instrument as well as the contribution to measurement error of technicians, the laboratory environment, and a laboratory's procedures. [IV.E]

52. a; The aim of calibration is to maintain a measuring system's accuracy. [IV.E]

53. b; In measurement systems analysis, there are two types of errors: *systematic* error and *random* error. Systematic error can be caused by humans, imprecise measuring devices, or poor manufacturing methods. It will remain relatively constant over repeated measurements. Random error is inherent fluctuations in the measurement system. Type I and type II errors are found in hypothesis testing (type I error occurs when a true null hypothesis is rejected; type II error occurs when a false null hypothesis is not rejected). [IV.F]

54. c; The accuracy of a measurement system consists of gage bias, linearity, and stability. Repeatability and reproducibility describe the precision of the measurement system. [IV.F]

55. a; Measurement system error can be attributed to gage bias, stability, linearity, repeatability, and reproducibility. Correlation measures the linear relationship between two quantitative variables. [IV.F]

56. a; *Bias* is defined as the observed average minus the reference value. In this example, the reference value is 6.00 cm. The observed average is

$$\bar{x} = \frac{6.05+5.94+5.98+6.01+6.03+5.97}{6} = 5.997.$$

Therefore, Bias = 5.997 – 6.00 = –0.003. [IV.F]

57. a; Consider the bias for each of the three types of parts. Recall that

Bias = Observed average – Reference value.

Part	1	2	3
Reference value	3.00	6.00	12.00
Observed average	2.998	5.843	12.275
Bias	–0.002	–0.157	0.275

Linearity measures how changes in the size of the part being measured affect the bias of the measurement system. In this example, the bias worsens as the length of the part increases, indicating evidence of nonlinearity in the measurement system. *Stability* is a measure of how well the measurement system performs over time. *Repeatability* is the variability of the test instrument when used to measure the *same* part. *Reproducibility* is the variability due to different operators or setups of the measurement system measuring the *same* part. [IV.F]

58. b; *Precision* is defined as the variability encountered when the same part is repeatedly measured using the same measurement system under identical conditions. Precision is made up of repeatability and reproducibility. *Accuracy* refers to the relationship between the results from a measurement system and the target or reference value. [IV.F]

59. c; Note that in this gage R&R study,

$$\sigma^2_{\text{Reproducibility}} = 0.8842 \text{ and } \sigma^2_{\text{Repeatability}} = 0.2466.$$

$$\sigma^2_{\text{Measurement error}} = \sigma^2_{\text{Reproducibility}} + \sigma^2_{\text{Repeatability}} = 0.8842 + 0.2466 = 1.1308.$$

[IV.F]

Part IV Solutions

60. c; The variance components can be found as follows:

Operators:

$$\hat{\sigma}_O^2 = \frac{MS_O - MS_{PO}}{pr} = \frac{0.033 - 0.867}{5 \times 3} = 0$$

(Set to 0, since this variance component is negative)

Part × Operator:

$$\hat{\sigma}_{PO}^2 = \frac{MS_{PO} - MS_E}{r} = \frac{0.867 - 1.233}{3} = 0$$

(Set to 0, since this variance component is negative)

Parts:

$$\hat{\sigma}_P^2 = \frac{MS_P - MS_{PO}}{or} = \frac{11 - 0.033}{2 \times 3} = 1.828$$

Error:

$$\hat{\sigma}_e^2 = MS_e = 1.233$$

$$\hat{\sigma}^2_{\text{Reproducibility}} = \hat{\sigma}_{PO}^2 + \hat{\sigma}_O^2 = 0$$

$$\hat{\sigma}^2_{\text{Repeatability}} = \hat{\sigma}_e^2 = 1.233$$

$$\hat{\sigma}^2_{\text{Measurement error}} = \hat{\sigma}^2_{\text{Gage}} = \hat{\sigma}^2_{\text{Reproducibility}} + \hat{\sigma}^2_{\text{Repeatability}} = 0 + 1.233 = 1.233$$

[IV.F]

61. d; Based on the *p*-values from the ANOVA table from question 60, there is not a significant operator effect, indicating that the operators are performing well and no additional training is needed. Also, from the ANOVA table, there is not a significant interaction between part and operator. However, by looking at the percentage contribution of the gage R&R variation to the total variability, the variability associated with repeatability is quite large. This indicates that the gage is not capable and the process needs to be investigated to reduce the variation associated with repeatability. [IV.F]

Source	Variance component	% contribution
Total gage R&R	1.233	40.28
Repeatability	1.233	40.28
Reproducibility	0	0
Operator	0	0
Operator × Part	0	0
Part-to-part	1.828	59.72
Total variation	**3.061**	**100**

62. a; From question 60,

$$\hat{\sigma}^2_{\text{Gage}} = \hat{\sigma}^2_{\text{Measurement error}} = 1.233.$$

Therefore,

$$\text{PTR} = \frac{6\hat{\sigma}_{\text{Gage}}}{\text{USL} - \text{LSL}} = \frac{6 \times \sqrt{1.233}}{30 - 10} = 0.333.$$

[IV.F]

63. b; An *R* chart provides information about special causes of variation related to the gage R&R study. In this example, no points plot outside the control limits, indicating that the operators behave similarly and that there are no special causes of variation (for example, operator fatigue or operator training issues). [IV.F]

Part V
Continuous Improvement

(60 questions)

QUESTIONS

1. A team at a cell phone manufacturer would like to reduce the number of defects in the manufacture of their cell phones. They identify potential defects of the cell phone. What quality control tool is most appropriate to determine the most frequent defects?

   a. Pareto chart

   b. Scatter diagram

   c. Flowchart

   d. Control chart

2. A team at a radiology clinic in a hospital wants to improve patient flow in their clinic. Before making changes to their process, the team discusses potential causes of disruptions in the patient flow. Which quality control tool will best help this team organize this information?

   a. Flowchart

   b. Scatter diagram

   c. Cause-and-effect diagram

   d. Check sheet

3. The team in the radiology clinic has collected data on patient visit times. The team suspects that the data are heavily skewed. Which of the following quality control tools would be most appropriate to confirm the team's suspicions?

   a. Histogram

   b. Scatter diagram

   c. Pareto chart

   d. Control chart

4. A team at a manufacturing facility is in the beginning stages of a performance improvement project to reduce the number of defects in a toy they produce. The team would like to collect data to decide the best opportunities for improvement. Which quality control tool is most appropriate to organize this data?

   a. Flowchart

   b. Check sheet

   c. Scatter diagram

   d. Histogram

5. A team at a bank wants to improve the time it takes to process a loan application. To identify any bottlenecks in the process or targets for improvement, which quality control tool is most appropriate to use?

   a. Pareto chart

   b. Check sheet

   c. Scatter diagram

   d. Flowchart

6. The bank has implemented changes to their loan application process and would like to monitor their improvements to ensure they are sustained. Which of the following quality control tools would best help this effort?

   a. Control chart

   b. Flowchart

   c. Check sheet

   d. Scatter diagram

7. A team suspects that two variables are related. Which of the following quality control tools will help them determine this?

   a. Histogram

   b. Flowchart

   c. Pareto chart

   d. Scatter diagram

8. A project team in a company comprises a large group of employees from different departments. The project has many complex components that depend on the progress of the various departments. Which of the following tools would help this team maintain a schedule for their project?

a. Prioritization matrix

b. Activity network diagram

c. Tree diagram

d. Interrelationship digraph

9. A clinic in a hospital is beginning a project to improve patient wait times. The project team is determining critical-to-quality factors. Which tool would be most appropriate to use?

   a. Value stream map

   b. Cause-and-effect diagram

   c. PVC diagram

   d. SIPOC diagram

10. A national print and copy company is considering changing the layout of their stores across the country to better serve their customers. The store is in the early phase of implementing their rollout plan for these changes. Which of the following tools would help the company evaluate their rollout plan to ensure there are minimal disruptions?

    a. Matrix diagram

    b. Process decision program chart

    c. Activity network diagram

    d. Prioritization matrix

11. An accident has recently occurred in a factory's shipping area. A team was put together to analyze potential actions that could lead to similar accidents. The goal is for the team to help determine the risk associated with these actions and to prevent future accidents from occurring. Which quality tool would help this team visualize this scenario?

    a. Fault tree

    b. Affinity diagram

    c. Matrix diagram

    d. Process decision program chart

12. A management team has a brainstorming meeting to identify opportunities to improve communication within their department. Which quality tool will best help them organize all the ideas generated by the team?

a. Tree diagram

b. Matrix diagram

c. Affinity diagram

d. Activity network diagram

13. Which of the following tools is common in quality function deployment and used to map customer requirements with technical specifications of a product?

a. Prioritization matrix

b. Activity network diagram

c. Matrix diagram

d. Interrelationship digraph

14. A medical team meets to discuss potential cause-and-effect relationships of actions that could lead to unfavorable outcomes during a patient surgical operation. The team consists of nurses, doctors, physician assistants, and technologists. Which of the following tools would best help this team visualize and hopefully identify factors that cause unfavorable outcomes?

a. Affinity diagram

b. Interrelationship digraph

c. Prioritization matrix

d. Activity network diagram

15. A cell phone manufacturer is developing a new cell phone. The research and development team has been given several required components of the phone, but have not decided on a design plan. Which of the following quality tools would help this team explore potential design plans for the new cell phone?

a. Fault tree

b. Prioritization matrix

c. Concept fan

d. Process decision program chart

16. The research and development team have now proposed several designs for the new cell phone. The team is now considering which of the designs to pursue to build a prototype. Which of the following tools would best help the team reach a decision on which of the designs to use?

a. Prioritization matrix

b. Affinity diagram

c. Interrelationship digraph

d. Process decision program chart

17. A team in the packing department of a shoe company noticed some inefficiencies in their packing process. The team first reduced the amount of walking employees had to do by moving the labeling station next to the packing station. The team continued to make incremental changes to their process, each of which led to a small improvement of the process. Which improvement technique has this team made use of?

a. Six Sigma

b. Total quality management (TQM)

c. Kaizen

d. Plan–do–check–act (PDCA)

18. In the PDCA cycle, in which stage of this improvement method does a team define the problem and develop an action plan?

a. P

b. D

c. C

d. A

19. Which of the following continuous improvement methods focuses on a business's customers, quality built into the work culture, and strong leadership involvement in quality efforts?

a. Kaizen

b. TQM

c. PDCA

d. Six Sigma

20. A quality team in a semiconductor factory, using the PDCA cycle, has proposed a change to their process. They have implemented the change and are currently analyzing the results. In which step of the PDCA cycle is this quality team?

a. P

b. D

c. C

d. A

21. What is the primary goal of kaizen improvement?
    a. Eliminate waste in a process
    b. Reduce the number of defects produced in a process
    c. Redesign a process to make large improvements in quality
    d. Make small, incremental changes to a process

22. What is one of the primary goals of Six Sigma as a continuous improvement method?
    a. Eliminate waste in a process
    b. Reduce the number of defects produced in a process
    c. Redesign a process to make large improvements in quality
    d. Make small, incremental changes to a process

23. What is the primary strength of theory of constraints?
    a. Incremental changes to a process can lead to great improvements.
    b. A system cannot improve by focusing improvements solely on individual processes.
    c. Team-based projects are more effective than individual-based projects.
    d. A data-driven improvement strategy is necessary for process improvement.

24. In Six Sigma, what value of sigma (σ), the standard deviation of a process, is used to account for typical shifts or drifts from the mean?
    a. $\pm 1.5\sigma$
    b. $\pm 2\sigma$
    c. $\pm 3\sigma$
    d. $\pm 6\sigma$

25. A restaurant with a particularly large kitchen and lots of empty space had long delays preparing and serving patrons. After careful planning, the restaurant completely redesigned their kitchen and ordering procedures, and immediately saw a vast decrease in serving times. They also noticed an increase in customers after this change. Which of the following improvement methods does this scenario describe?
    a. PDCA
    b. TQM

c. Reengineering

d. Kaizen

26. What value of C_p is associated with a Six Sigma–quality level process?

a. 1.33

b. 1.5

c. 2.0

d. 2.5

27. Which of the following methods is the most effective way to implement Six Sigma in an organization?

a. Start a Six Sigma program.

b. Develop a Six Sigma infrastructure.

c. Roll out Six Sigma methods on a project-by-project basis.

d. Begin at lower levels in the organization.

28. Which of the following tools is typically used to identify waste in a lean system?

a. 5S

b. Kanban

c. Value stream map

d. Visual control

29. A restaurant is analyzing its current workflow. Employees in the restaurant must walk back and forth throughout the kitchen to retrieve various cooking tools during preparation of a meal. How might the restaurant classify this action?

a. Waiting

b. Excess motion

c. Excess movement of material

d. Excess processing

30. The restaurant is also examining its front-of-house operations. They found that there is frequently a long line of customers standing by the chicken counter while staff turn over tables. What kind of waste is occurring at the front of house at this restaurant?

a. Waiting

b. Excess movement of material

c. Defect correction

d. Excess motion

31. After creating a value stream map of a process, on which steps in the process should a quality team focus their efforts?

    a. Value-added activities

    b. Standardized work

    c. Non-value-added activities

    d. Transportation of the product

32. What is the primary goal of 5S?

    a. To radically change a process

    b. To make slow, incremental changes to a process

    c. To organize the workplace environment

    d. To reduce defects produced in a process

33. A toy company paints a certain toy white by default. The toys are then painted other colors as requested by customers. As what type of waste might a quality team classify this action?

    a. Defect correction

    b. Excess motion

    c. Excess processing

    d. Excess production

34. What is the main benefit of implementing a kanban system?

    a. The weakest step in a chain of processes is improved.

    b. There is less inventory and a smoother flow of supplies.

    c. The workplace is better organized.

    d. Products are not processed until a customer places an order.

35. A deli has newly instituted a few changes to its work area. The workers now have a schedule for cleaning their workstations, have standardized practices, and have created designated locations for all their equipment. What improvement method has this deli employed?

    a. Kanban

    b. Visual control

c. Define, measure, analyze, improve, control (DMAIC)

d. 5S

36. A factory uses an illuminated indicator light when a certain process is in progress. This is an example of:

    a. waste.

    b. standardized work.

    c. visual control.

    d. kanban.

37. In a convenience store, items are stacked in front of a card that lists details on the product, including the location of additional boxes in their storage room as well as a number to call to reorder. What type of system has this convenience store implemented?

    a. Kaizen

    b. 5S

    c. TQM

    d. Kanban

38. To complete a process, parts must travel through four departments located in different areas of a manufacturing facility. As what type of waste would a quality team at the facility classify this process?

    a. Excess movement of material

    b. Defect correction

    c. Excess processing

    d. Excess motion

39. A manufacturing facility has painted arrows on the floor to show the flow of products through the facility. Which lean tool has this facility implemented?

    a. Kanban

    b. Visual control

    c. 5S

    d. Kaizen

40. The principle that each process should be done the same way, every time, by every employee is called:

    a. kanban.

    b. 5S.

    c. standardized work.

    d. pull system.

41. Which of the following lean tools measures the time from the last good part of one type to the first good part of the next?

    a. Takt time

    b. Single-minute exchange of die (SMED)

    c. Cycle time

    d. 5S

42. What is the difference between cycle time and takt time?

    a. Cycle time incorporates data from suppliers; takt time does not.

    b. Takt time is based on customer demand; cycle time measures the time to complete an activity.

    c. Takt time measures the time to complete an activity; cycle time incorporates customer demand.

    d. There is no difference; they both measure the same time.

43. A shoe company routinely makes more components than are needed when assembling its product. As what type of waste might this company classify this action?

    a. Inventory

    b. Overproduction

    c. Excess movement of material

    d. Excess processing

44. A university has released a document detailing all the required actions instructors should take when proctoring an exam. All instructors across the university are expected to follow these directions. What tool has the university implemented?

    a. Standardized work

    b. 5S

c. Visual control

d. Value stream map

45. A company has received an order for 550 items that it needs to produce in an eight-hour workday. In seconds, what is the takt time for this product?

a. 0.87 seconds

b. 52.36 seconds

c. 68.75 seconds

d. 550 seconds

46. A caterer had a lunch order for a group of 50 people and had two hours to complete the order. However, after two hours, the caterer was only able to complete 45 lunches. Which of the following must be true?

a. Cycle time ≤ Takt time

b. Cycle time > Takt time

c. Cycle time = Takt time

d. Not enough information to determine

47. A radiation clinic treats patients with different types of cancer. The position of each patient on the table is different depending on the type of cancer. Therefore, after one patient is treated, another patient, requiring a different setup than the previous patient, arrives. Which lean tool would best help a quality team in this clinic?

a. Cycle time

b. 5S

c. SMED

d. Kanban

48. A supplier has received an order from one of its customers. It must complete an order of 750 items in a daytime shift that is 10 hours long. The cycle time for this process is 50 seconds. Without any changes in its current production, will the supplier be able to meet the customer's demands?

a. No, since cycle time ≤ takt time.

b. Yes, since cycle time ≤ takt time.

c. No, since cycle time > takt time.

d. Yes, since cycle time > takt time.

49. Which of the following is a speed loss in OEE?

    a. Breakdown loss

    b. Process defect

    c. Yield loss

    d. Idling and minor stoppage loss

50. If an eight-hour shift corresponds to 480 minutes, what is the availability of a machine with the following information:

    - There are two breaks of 15 minutes each.
    - The machine broke down for 35 minutes.
    - The adjustment time was 10 minutes.

    a. 82%

    b. 90%

    c. 91%

    d. 95%

51. What is one of the most common causes of failures in problem solving?

    a. Poor definition of the problem

    b. Multifunctional teams

    c. Selecting projects that require the least amount of effort

    d. The project charter

52. During the correction phase of the corrective action cycle, which of the following is instrumental in success?

    a. A clear problem statement

    b. Team meetings

    c. Creativity

    d. Cause-and-effect diagram

53. A team in a food production plant has recently instituted changes to one of their bottling processes. What should the team do to avoid the process backsliding?

    a. Identify key metrics in the process to monitor

    b. Ensure management can perform each task in the process

c. Remove process certification for operators

d. Set production quotas

54. In which phase of the corrective action cycle does a team seek to continuously improve the performance of a process by holding regular reviews of the process?

    a. Problem identification

    b. Correction

    c. Recurrence control

    d. Effectiveness assessment

55. A 5 Why Analysis is complete when:

    a. the fifth "Why" has been completed.

    b. the root cause has been identified.

    c. the cause-and-effect diagram has been completed.

    d. the brainstorming session has been completed.

56. When utilizing a 5 Why Analysis to solve a manufacturing problem, which employee would be best suited to have firsthand knowledge of the system in place?

    a. Design engineer

    b. Calibration technician

    c. Purchasing representative

    d. Production operator

57. Which of the following is another name for poka-yoke?

    a. Waste

    b. Error-proofing

    c. Orderliness

    d. Pull system

58. A food company produces cookie dough. All a customer must do is slice the cookie dough and bake it in their oven for approximately eight minutes. The company's cookie dough recipe was developed so that the cookie quality is not vastly affected by differences in ovens, elevation, or other environmental conditions. This cookie dough is an example of

    a. poka-yoke.

b. reengineering.

c. robust design.

d. kaizen.

59. A printer that will only run when the paper trays are closed is an example of which of the following concepts?

a. Robust design

b. Poka-yoke

c. Lean process

d. Kaizen

60. Before the next step in a process, a sorting machine is used to remove components that are above the upper specification limit. What principle of error-proofing is implemented in this process?

a. Elimination

b. Replacement

c. Facilitation

d. Detection

SOLUTIONS

1. a; Pareto charts are used to identify the most frequent issues in a process. The chart displays the "80/20 rule"—approximately 20% of factors (defects) account for approximately 80% of potential issues. Scatter diagrams graphically compare the relationship between two continuous variables, flowcharts provide a graphical representation of a process, and control charts are used to identify out-of-control behavior in a process. [V.A]

2. c; Cause-and-effect diagrams (also called *fishbone* or *Ishikawa* diagrams) are used to identify all major and minor factors that contribute to a problem. These diagrams can help the team identify the best opportunities for improvement in patient flow in the clinic. [V.A]

3. a; Histograms are a graphical representation of the frequency of occurrences of some event. They provide a visual depiction of the shape and spread of the data and can therefore help the team decide whether the data are skewed, symmetric, bimodal, and so on. Scatter diagrams are used to depict the relationship between

two continuous variables. Pareto charts are graphical tools used to identify the most frequent events. [V.A]

4. b; Check sheets are useful tools for summarizing counts of events, for example, the count of different types of defects. The data collected from a check sheet can then be graphically represented with a Pareto chart to identify the most frequent defects. Scatter diagrams graphically depict the relationship between two continuous variables, and histograms provide a graphical representation of the distribution for a continuous variable. [V.A]

5. d; Flowcharts are a graphical representation of a process and can help identify areas for improvement (like a bottleneck). Pareto charts are used to identify the most common defects, check sheets are used to tally occurrences of events, and scatter diagrams are used to graphically depict the relationship between two continuous variables. [V.A]

6. a; Control charts can be used to monitor a quality characteristic, in this case the time until the loan applications are processed. A time that plots beyond the control limits will indicate that a special cause has occurred. This alert will help the team correct the problem and maintain their improvements. [V.A]

7. d; Scatter diagrams are used to graphically depict the relationship (if there is one) between two continuous variables. Histograms are graphical tools used to depict the distribution of one continuous variable. [V.A]

8. b; Activity network diagrams help to organize complex schedules or sequences of events. These diagrams, in particular, can help with time scheduling. Because this project involves a large group of people and is dependent on the work of different departments, an activity network diagram would help keep all the team members aware of the project schedule to avoid bottlenecks during the project. [V.B]

9. d; A SIPOC diagram is a process map that contains information on the suppliers, inputs, outputs, and customers of a process. The diagram helps to capture the voice of the customer and the critical-to-quality factors. [V.B]

10. b; A process decision program chart (PDPC) would help this company assess its national rollout plan at a high level, since they are at the beginning of their implementation plan. These charts help to determine potential disruptions to a process plan so that the company can prepare for any necessary contingency plans. [V.B]

11. a; Fault trees provide a visualization of hierarchical relationships of events that lead to some failure (or other undesirable final outcome). The team here wishes to determine other potential actions that can lead to a similar accident in the factory.

The fault tree would help the team identify and communicate other potential causes of the accident. The fault tree would also provide a means to quantify the risks of these potential causes. [V.B]

12. c; Affinity diagrams are used to help a group of people organize or summarize facts, opinions, or ideas. Often, these types of diagrams are built using sticky notes that are grouped into common themes or categories. [V.B]

13. c; Matrix diagrams are used to display qualitative relationships between two sets of characteristics or factors. Therefore, these diagrams are common in quality function deployment because they clearly display the relationships between customer requirements and a product's technical specifications. [V.B]

14. b; Interrelationship digraphs provide a means to identify sequential or cause-and-effect relationships of a problem or situation, in this case an unfavorable outcome in the operating room. Affinity diagrams are used to organize general thoughts and ideas about a topic and are more appropriate for grouping related ideas rather than discovering causal relationships. Prioritization matrices provide a formal method for comparing decisions or ideas before making a final decision. Activity network diagrams help to organize events for time schedule planning. [V.B]

15. c; A concept fan is a type of tree diagram that allows a team to explore alternate concepts that achieve the same goals or requirements. Fault trees are also tree diagrams but provide information on the hierarchical events that lead to an unfavorable outcome or failure in a process. Prioritization matrices provide a formal method for comparing decisions or ideas before making a final decision. Process decision program charts (PDPCs) help to evaluate a process implementation plan at a high level. [V.B]

16. a; Prioritization matrices are useful in the decision phase since they allow us to make relative comparisons of different ideas or concepts with consistent criteria. The R&D team has developed several competing designs and can use a prioritization matrix to fairly compare the designs to help them make a final decision. [V.B]

17. c; The packing team has made several minor changes to their process, each of which led to a small improvement. This describes kaizen, a methodology that encourages employees to make incremental changes continually to improve a process. Each change may not make a large improvement, but does help employees. [V.C]

18. a; In the *plan* phase of the plan–do–check–act (PDCA) cycle, the quality team decides on a potential change to the process. In the *do* phase, the team performs the proposed change, in the *check* step, the team analyzes the results of the

change, and in the *act* phase, the team decides either to accept the change or not, depending on whether the change was successful or not. [V.C]

19. b; Total quality management (TQM) is an organization-wide quality program that focuses on customers' involvement in determining quality, employees' obligation to quality, and strong leadership from upper management in quality improvement efforts. Kaizen is a method of continuous improvement that makes incremental changes to a process. PDCA, also called the Shewhart or Deming (PDSA) cycle, is a methodology for a performance improvement project, and theory of constraints (TOC) is a problem-solving methodology that focuses on the weakest part of a process. [V.C]

20. c; The quality team has already implemented the proposed change and are analyzing the results. This describes the *check* step of the PDCA cycle. In this step, the quality team should carefully analyze the results of their change to determine whether the change was successful or not. In the *act* step, the team will make a decision on whether to keep the change, and then repeat the cycle. [V.C]

21. d; The primary goal of kaizen is for employees to make continual, incremental changes to a process to achieve small improvements. Lean focuses on eliminating waste in a process, one of the goals of Six Sigma is to reduce defects, and reengineering involves redesigning a process to see large improvements in quality of a product. [V.C]

22. b; One of the primary goals of Six Sigma is to reduce the number of defects produced in a process. Lean focuses on eliminating waste in a process, reengineering involves redesigning a process to see large improvements, and kaizen is continuous, incremental improvements to a process. [V.C]

23. b; TOC focuses on the weakest link in a chain of processes—therefore, it's a systems approach to process improvement. Focusing on an individual process in a chain of many will not provide the best result. [V.C]

24. a; The sigma quality level includes a ±1.5< value to account for typical drifts from the mean. [V.C]

25. c; Reengineering a process involves a complete and radical redesign of a procedure. In this case, the restaurant has completely redesigned its kitchen and ordering procedure to achieve large improvements in its serving time and sales—an example of reengineering. [V.C]

26. c; Recall that the process capability index C_p is defined as

$$C_p = \frac{USL - LSL}{6\sigma}.$$

Since a three-sigma-quality level process has a C_p value of 1.0, a Six-Sigma–quality level process has a C_p value of 2.0. There would be twelve standard deviations total between the LSL and USL. [V.C]

27. b; One of the drawbacks of a Six Sigma program is that it often lacks buy-in across the entire organization because it focuses on the tools only. Therefore, results tend to be very limited. A Six Sigma infrastructure provides employees, not just with the tools, but also with the problem-solving methods for specific projects identified by management. [V.C]

28. c; The goal of a lean system is to eliminate waste from a process. One of the first steps is to identify opportunities for improvement, which can be found using a value stream map. Value stream maps are a visual representation of all the steps in a process, including information about timing. Value stream maps also identify the value-added and non-value-added activities in a process. [V.D]

29. b; The employees walk more than they should to retrieve supplies. This is an example of waste—specifically, that of excess motion. A redesign of the kitchen or a new layout that moves the cooking tools closer to where they are required could help reduce the excess walking and shorten the time it takes to prepare the meal. [V.D]

30. a; The customers are waiting to be seated at the restaurant due to the long setup/turnover times of the tables. The restaurant may want to cross-train their employees or introduce standardized work to help reduce the setup time and therefore the customer waiting time. [V.D]

31. c; Non-value-added activities are any steps in a process that are unneeded in the process. Improvement projects will have the largest impact on these non-value-added steps in a process. [V.D]

32. c; 5S stands for five Japanese words that detail ways to better organize a workplace. NIST defines these in English as sort, set in order, shine, standardize, and sustain. Note that reengineering involves radically changing a process, kaizen implements slow, incremental changes to a process, and one of the goals of Six Sigma is to reduce the number of defective items. [V.D]

33. c; This is an example of waste due to excess processing. The toys are all painted one color (white), then may be repainted per customer demands. The company may want to implement a new procedure for painting the toys so that they are not painted until a customer has requested their desired color. This would eliminate the need to paint the toys more than once. [V.D]

34. b; A kanban system provides visual cues to help ease restocking efforts. As a result, there is often less inventory and a better flow of supplies. The Theory of constraints (TOC) focuses on improving the weakest step in a chain of processes.

5S is used to organize a workplace. A pull system means that the steps in a process are completed only when triggered by a previous action, for example, a customer order. [V.D]

35. d; 5S stands for five words adopted from Japanese that details ways to better organize a workplace. These include arranging things properly, orderliness, cleanliness, cleanup, and discipline. By incorporating a system for cleaning, developing standardized practices, and maintaining order with their equipment, this deli is implementing the five S's. [V.D]

36. c; A visible signal, like a light, is an example of visual control. The light is a cue to the employees that an action is being performed. These visuals help make employees more aware of their surroundings, which can help create a safer or more efficient workplace. [V.D]

37. d; A kanban system uses visual cues to help resupply efforts. When a bin of a particular item at the convenience store is emptied, the card alerts the convenience store employee where to find additional supplies, or how to order more. [V.D]

38. a; This is an example of waste due to excess movement of material. A layout change in the facility that groups together all assembly processes for a given product could reduce the distance the components must travel. Note that excess motion refers to more motion than is necessary that employees or operators must make to complete an activity. [V.D]

39. b; This is an example of visual control. Arrows on the floor of the facility provide employees quick, visual cues on the flow of traffic. This will help movement in the facility and reduce errors and/or accidents. [V.D]

40. c; Standardized work means that every step or activity should be done the same way. Standardized work helps to reduce process variation and provides a more consistent product or service. [V.D]

41. b; Single-minute exchange of die (SMED) is a method used to reduce changeover times of machines. The goal of using SMED is to reduce the time required to switch a machine or assembly to a different product. [V.D]

42. b; Cycle time is the time it takes to complete an activity or process step. Takt time is based on customer demand and is defined as

$$\text{Takt time} = (\text{Time available}) \div (\text{Units required}).$$

The takt time therefore is the average time a process takes to complete one item out of the customer demand. [V.D]

43. b; This is an example of waste due to overproduction. This form of waste involves making more of a product than is necessary. In this case, the shoe company

consistently makes more components than it needs, which is a waste of resources and time. The company should adjust its production of this component. [V.D]

44. a; The university has implemented standardized work. All instructors are expected to follow the exam instructions the same way, every time. The goal of standardizing the exam procedure should help the process run more smoothly and make it more efficient across the university. [V.D]

45. b; Takt time is defined as

$$\text{Takt time} = (\text{Time available}) \div (\text{Units required}).$$

The time available in seconds is 8 × 60 × 60 = 28,800 seconds. Therefore,

$$\text{takt time} = 28{,}800 \div 550 = 52.36 \text{ seconds.}$$

Therefore, the company must average one item every 52.36 seconds. [V.D]

46. b; The takt time for the caterer (in minutes) is 2 × 60 ÷ 50 = 2.4 minutes. Therefore, the caterer needs to average a lunch every 2.4 minutes to meet customer demand. Since the caterer was only able to finish 45 lunches in the two hours, the cycle time must have been greater than the takt time. [V.D]

47. c; SMED is a method used to reduce changeover times. Although initially introduced in a factory setting, the same methodology could apply in the radiation clinic, as there are required changeover times for the different types of patients. [V.D]

48. c; The takt time for this system is 10 × 60 × 60 ÷ 750 = 48 seconds. The supplier must average one item every 48 seconds. Because the cycle time (50 seconds) is greater than the takt time, the supplier is unlikely to meet the customer demand without any adjustments to their production. [V.D]

49. d; The six big losses are listed in *The ASQ Certified Quality Engineer Handbook*, Fifth Edition. The speed losses are (1) idling and minor stoppage and (2) reduced speed. [V.D]

50. b; The loading time is calculated by subtracting the planned downtime from the total time. This is equal to 450 minutes (480 – 15 – 15). The unplanned downtime is calculated by adding all the unplanned downtime. This is equal to 45 minutes (35 + 10). Using these numbers:

$$\begin{gathered}\text{Availability} = (\text{Operation Time}/\text{Loading Time}) \times 100 \\ = ((\text{Loading Time} - \text{Unplanned Downtime}/\text{Loading Time}) \times 100 \\ = (450 - 45)/450) \times 100 = 90\%.\end{gathered}$$

[V.D]

Part V Solutions

51. a; One of the most frequent sources of failure in problem-solving is having a poor definition of the problem. Problem statements should be clear, focused, use facts and not judgment, and establish what is known, unknown, and needs to be done. [V.E]

52. c; In phase one of the corrective action cycle, the problem is identified and a clear problem statement is required. In phase two, the current process is analyzed and the root cause is identified, frequently through cause-and-effect diagrams. In the next phase, correction, an optimal solution is developed to solve the problem. In this phase, creativity is essential to best solve the problem. [V.E]

53. a; The recurrence control phase of the corrective action cycle involves the standardization of the solution to the problem. In this phase, the team should identify the important variables that need to be kept under control. [V.E]

54. d; In the problem identification phase, the team identifies areas for improvement and develops a clear problem statement. In the correction phase, the team develops feasible solutions to the problem and recommends the best choice to implement. In the recurrence control phase, the team standardizes the solution to ensure that the problem does not reoccur and prevent backsliding. Finally, in the effectiveness assessment phase, the team continues monitoring the process to identify additional opportunities for improvement. [V.E]

55. b; As stated in *The ASQ Certified Quality Engineer Handbook*, Fifth Edition: "Loop back to step 3 until the team agrees that the problem's root cause is identified. Again, this may take fewer or more times than five Why's." Keep in mind that "5" is just a number. Ask "Why" as many times as needed to complete the process and take appropriate actions. When the root cause of the problem is revealed, asking "why" again produces no more useful responses, and there is no need to go no further. [V.E]

56. d; When solving a manufacturing problem, the most appropriate department would be manufacturing, of which a "Production Operator" would be an employee. As a member of the manufacturing department, the Production Operator would be the closest and therefore have the most firsthand knowledge of the systems in place. [V.E]

57. b; Poka-yoke comes from Japan and is also known as error-proofing. It is a method of preventive action—a technique used to prevent errors from occurring. Note that waste is also known as *muda*, and orderliness is one of the five S's (*seiton*). [V.F]

58. c; Because the quality of the cookie is not greatly affected by differing use conditions, this is an example of robust design. A robust design prevents or reduces variation in quality under various use conditions. [V.F]

59. b; Poka-yoke, or error-proofing, is used for error prevention. The printer not being allowed to run if a paper tray is open prevents paper jams or other possible errors. [V.F]

60. d; Detection is the principle that a defect or error is caught before it moves to the next step in the process. In this scenario, the sorting machine removes components that exceed the upper specification limit before moving to the next step in the process. [V.F]

Part VI
Quantitative Methods and Tools

(115 questions)

QUESTIONS

1. A project is being conducted in a hospital clinic. Which of the following are considered continuous data?
   a. Patient weight
   b. Result of a pass/fail medical test
   c. Errors in a medical record
   d. Number of patient visits
2. A project is being conducted in a manufacturing facility. Which of the following are considered discrete data?
   a. The length of a component
   b. Fill volume of a container
   c. Number of nonconforming components
   d. Time required for testing
3. The color of a car is on which measurement scale?
   a. Nominal
   b. Ordinal
   c. Interval
   d. Ratio
4. A person's height in feet is on which measurement scale?
   a. Nominal
   b. Ordinal
   c. Interval
   d. Ratio

5. The rating of a service from poor to excellent is on which measurement scale?
   a. Nominal
   b. Ordinal
   c. Interval
   d. Ratio
6. The Data Transform phase's main goal is to
   a. remove outliers.
   b. normalize data.
   c. prepare data to fit the parameters of another system.
   d. split or join tables.
7. Which of the following distributions best describes the data set below?

Goodness of Fit Test Results

Normal AD = 1.456 $p < 0.005$	Weibull AD = 0.214 $p > 0.250$
Exponential AD = 1.456 $p < 0.003$	Gamma AD = 0.581 $p = 0.152$

   a. Normal
   b. Exponential
   c. Weibull
   d. Gamma
8. Which of the following can simplify recording and analysis of data?
   a. Regression
   b. Coding
   c. DOE
   d. Matrix plot
9. The waiting times in minutes for eight customers at a bank are: 5, 8, 12, 3, 2, 7, 6, 5. What are the mean and median for this sample of waiting times?
   a. $\overline{X} = 5.5$ minutes, M = 6 minutes
   b. $\overline{X} = 6$ minutes, M = 5 minutes

c. $\bar{X} = 6$ minutes, M = 5.5 minutes

d. $\bar{X} = 6$ minutes, M = 6 minutes

10. The lengths in inches for a sample of seven components are: 3.2, 3.1, 3.4, 3.1, 3.2, 3.3, 3.2. What is the standard deviation for this sample?

    a. $s = 0.0098$ inches

    b. $s = 0.0989$ inches

    c. $s = 0.0114$ inches

    d. $s = 0.1069$ inches

11. The median of a population is much larger than the population mean. What can you say about the shape of the population distribution?

    a. The distribution is symmetric.

    b. The distribution is right-skewed.

    c. The distribution is left-skewed.

    d. The distribution has high kurtosis.

12. A population has an exponential distribution. One hundred samples of size 40 are randomly collected and the 100 sample means are calculated. What is the approximate distribution of these sample averages?

    a. Binomial

    b. Normal

    c. Uniform

    d. Poisson

13. What is the formula for the population variance?

    a. $\dfrac{\Sigma_{i=1}^{N}(x_i - \mu)}{N}$

    b. $\dfrac{\Sigma_{i=1}^{n}(x_i - \bar{x})^2}{n-1}$

    c. $\dfrac{\Sigma_{i=1}^{N}(x_i - \mu)^2}{N}$

    d. $\dfrac{\Sigma_{i=1}^{N}(x_i - \mu)^2}{N-1}$

14. A team would like to compare waiting times of customers by day of the week. Which of the following would be the most appropriate tool to use?

    a. Histograms

    b. Box plots

    c. Scatter plots

    d. Probability plots

15. Which of the following descriptive statistics are typically found in a box plot?

    a. Minimum and maximum

    b. Minimum and mean

    c. Maximum and mean

    d. Median and mean

16. A team measured the time to failure of twenty components. They plotted the data in a normal probability plot, shown below. What conclusions can be drawn from this plot?

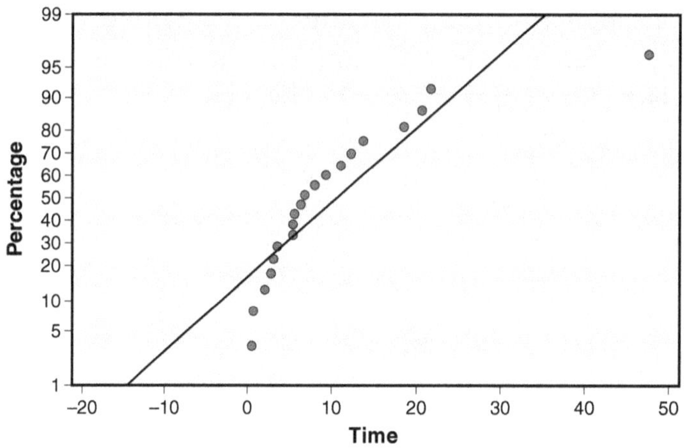

    a. The normal distribution is not a reasonable model for the time to failure.

    b. The Weibull distribution is a reasonable model for the time to failure.

    c. The normal distribution is a reasonable model for the time to failure.

    d. Not enough information is provided to draw conclusions.

17. The temperature of coffee is an important characteristic in a restaurant. The temperature of fifteen randomly selected cups of coffee is measured and the average temperature is found to be 168°F. Which term best describes this average temperature?

a. Population

b. Sample

c. Parameter

d. Statistic

18. A characteristic of a population is called a:

a. sample.

b. statistic.

c. parameter.

d. probability.

19. The contingency table below describes the relationship between the number of nonconforming components produced and time of day.

Time	Conforming parts	Nonconforming parts	Totals
Day	94	5	99
Evening	87	10	97
Overnight	65	23	88
Totals	246	38	284

Given that a part is made overnight, what is the probability it is nonconforming?

a. 0.08

b. 0.26

c. 0.43

d. 0.61

20. The probability of a part being defective is 0.02. What is the probability it is not defective?

a. 0.02

b. 0.88

c. 0.98

d. 1.08

21. The probability that event A occurs is 0.72. The probability that event B occurs is 0.03. What is the probability that events A and B both occur?

    a. 0.0216

    b. 0.69

    c. 0.75

    d. Not enough information is provided.

22. If the probability that event A occurs is 0.40, the probability that event B occurs is 0.13, and A and B are mutually exclusive events, what is the probability that events A and B both occur?

    a. 0

    b. 0.052

    c. 0.53

    d. 1.27

23. The time between arrivals at a store has an exponential distribution with mean five minutes. What is the probability that the time between arrivals is less than three minutes?

    a. 0

    b. 0.4512

    c. 0.5488

    d. 0.9999

24. X has a normal distribution with mean 15 and standard deviation 2. What is the probability that X is greater than 18?

    a. 0.0668

    b. 0.1446

    c. 0.8554

    d. 0.9332

25. Which of the following probability distributions is asymmetric?

    a. Normal

    b. *t*

    c. Uniform

    d. Weibull

26. A bottling company's filling process has a normal distribution with mean 24.01 oz. and standard deviation 0.025 oz. If the specifications for the process are 24 ± .08 oz, what proportion of bottles is overfilled?

    a. 0.00016

    b. 0.00256

    c. 0.99488

    d. 0.99744

27. Compared to the standard normal distribution, the student's *t*-distribution:

    a. has a larger mean.

    b. is more skewed to the right.

    c. is bimodal.

    d. has heavier tails.

28. The number of nonconformities per inspection unit in a factory occurs at a rate of five per day. Using the most appropriate probability distribution, what is the probability there will be at least two nonconforming units in one day?

    a. 0.0404

    b. 0.2627

    c. 0.8000

    d. 0.9596

29. A sample of ten products is selected from a small, isolated lot without replacement. The lot is known to contain 5% defective products. Which distribution is most appropriate to determine the probability of selecting at least one defective item in the sample?

    a. Hypergeometric

    b. Binomial

    c. Poisson

    d. Normal

30. Fifteen invoices are randomly selected from a batch at a company. It is believed that 10% of all invoices have an error. What is the probability that exactly four out of the fifteen invoices have an error?

    a. 0.0127

    b. 0.0428

c. 0.9873

d. 1.0421

31. Suppose the number of errors in an invoice at a company has the probability mass function (written in table form) shown below. What is the expected number of errors in a randomly selected invoice?

Number of errors X	0	1	2	3	4
Probability	0.86	0.06	0.04	0.03	0.01

a. 0

b. 0.27

c. 0.5771

d. 0.7597

32. A population has a distribution with mean 24 and standard deviation 5.5. According to the central limit theorem, what is the mean of the distribution of the sample means for random samples of size 32?

a. 24/32

b. $24/\sqrt{32}$

c. 24/5.5

d. 24

33. A bottling company's filling process has a normal distribution with mean 24 oz. and standard deviation 0.05 oz. What is the probability that a sample of sixteen randomly selected bottles will have an average volume more than 24.04 oz?

a. 0

b. 0.0007

c. 0.2119

d. 9993

34. A sample of size 16 is drawn from a population that follows a normal distribution with mean 58 and variance 64. What is the standard error of the sample mean?

a. 2

b. 4

c. 8

d. 16

35. The 95% confidence interval for the proportion of filing errors in a business is (0.02, 0.12). What conclusions can be drawn from this confidence interval?

   a. There is a 5% chance that the population proportion p is between 0.02 and 0.12.

   b. The probability that the confidence interval contains the population proportion p is 0.5.

   c. The error rate is unacceptable.

   d. There is 95% confidence that the confidence interval will contain the population proportion p.

36. A team would like to estimate the average length of a component. A random sample of 50 components had a sample mean of 11.75 mm. Which of the following terms is used to describe this value?

   a. Point estimate

   b. Parameter

   c. Probability

   d. Standard error

37. The tensile strength of fifteen units of cement was tested. The sample mean tensile strength was 4.3 MPa and sample standard deviation was 0.51 MPa. Find a 99% confidence interval for the true mean tensile strength of the cement.

   a. (2.782, 5.818)

   b. (3.908, 4.692)

   c. (3.961, 4.639)

   d. (4.018, 4.582)

38. A team would like to estimate the true average tensile strength of cement. They would like to obtain an estimate within 0.15 MPa of the true average tensile strength with 95% confidence. Based on prior information, it is assumed that $\sigma = 0.50$ MPa. What sample size is required to meet these requirements?

   a. 6

   b. 47

c. 42

d. 43

39. The voltage of a power supply is of interest to a team. Voltage is assumed to be normally distributed. The voltages of seven randomly selected observations are: 10.37, 11.50, 9.80, 10.65, 11.95, 10.15, 9.52. Find the 95% confidence interval on the population variance of voltage.

    a. (0.324, 3.788)

    b. (0.378, 4.419)

    c. (0.570, 1.946)

    d. (9.745, 11.380)

40. A bottling factory claims that the volume of soda filled in bottles is 18 oz. You want to test whether the true mean volume of soda is different than 18 oz. The volume of a random sample of twenty-four bottles had a mean of 17.98 oz. with a standard deviation of 0.03 oz. Using a significance level of 5%, what is the appropriate test statistic for this hypothesis test?

    a. –3.266

    b. 0.607

    c. 2.069

    d. 3.266

41. Suppose that the *p*-value of the hypothesis test posed in the previous question (question 40) is 0.003. What conclusions can be made (use significance level 5%)?

    a. There is not sufficient evidence to conclude that the mean volume is different than 18 oz.

    b. There is sufficient evidence to conclude that the mean volume is different than 18 oz.

    c. 18 would be contained in a 95% confidence interval of the population mean.

    d. 18 would be contained in a 90% confidence interval of the population mean.

42. Speed bumps were installed in a neighborhood to slow traffic. After installation, the speeds after the last speed bump of a random sample of twenty cars were recorded. The mean speed was 24.50 mph and the standard deviation was 2.3 mph. The team wanted to determine if the average speed is less than 25 mph. What are the appropriate hypotheses for this hypothesis test?

    a. $H_0 : \mu = 25$ versus $H_a : \mu \neq 25$

b. $H_0 : \mu > 25$ versus $H_a : \mu = 25$

c. $H_0 : \mu = 25$ versus $H_a : \mu < 25$

d. $H_0 : \mu = 25$ versus $H_a : \mu > 25$

43. A team investigated the number of patient falls in a hospital. Fifty-six patients were randomly selected. Out of the fifty-six patients, there were seven recorded patient falls. The hospital wanted to determine whether the proportion of patient falls was more than 10%. At the 5% significance level, what is your conclusion about this hypothesis test?

a. Reject H_0, and conclude that the proportion of patient falls equals 0.10.

b. Reject H_0, and conclude that the proportion of patient falls is more than 0.10.

c. Do not reject H_0; conclude that the proportion of patient falls equals 0.10.

d. Do not reject H_0; conclude that the proportion of patient falls is more than 0.10.

44. Two methods for filling a bottle are being compared at a factory. The 90% confidence interval for the difference in volume for two independent random samples of the filling methods was found to be (–0.091, –0.049). Based on this confidence interval, is there evidence that the two filling methods are different?

a. Yes, since 0 is not in the confidence interval.

b. Yes, since 0 is in the confidence interval.

c. No, since 0 is not in the confidence interval.

d. No, since 0 is in the confidence interval.

45. An improvement project was done at a factory to decrease packaging time. The following data were collected before and after the improvement project:

Before: Sample mean = 3.52, Sample variance = 2.34, Sample size = 28

After: Sample mean = 2.10, Sample variance = 2.04, Sample size = 25

A hypothesis test was performed with $H_0 : \mu_{Before} - \mu_{After} = 0$ versus $H_a : \mu_{Before} - \mu_{After} > 0$. What is the critical value for this hypothesis test (Use $\alpha = 0.01$)?

a. –2.402

b. 2.402

c. 2.576

d. 2.678

46. Consider the scenario in the previous problem (question 45). The project team would like to determine whether the variance of packaging time has changed at the 10% significance level. What is the test statistic for the appropriate hypothesis test?

    a. 1.071

    b. 1.147

    c. 1.930

    d. 3.494

47. Two machines produce the same parts. A random sample of 1,250 parts from machine 1 has twenty-eight that are nonconforming, and a random sample of 1,175 parts from machine 2 has eighteen that are nonconforming. Find the 90% confidence interval for the difference between the proportions of nonconforming parts from machines 1 and 2.

    a. (–0.004, 0.018)

    b. (–0.002, 0.016)

    c. (0, 0.016)

    d. (0.006, 0.008)

48. A hypothesis test is performed at the 10% significance level. The power of the test is 0.80, or 80%. What is the type II error for this hypothesis test?

    a. 0.10

    b. 0.20

    c. 0.80

    d. 0.90

49. When the null hypothesis is not rejected when in fact the null hypothesis is false, what type of error has been made?

    a. Sampling error

    b. Type I error

    c. Type II error

    d. No error has been made

50. A team would like to compare the proportion of defective widgets produced by two machines. What distribution does the test statistic for this hypothesis test follow?

a. Normal

b. t

c. Poisson

d. Uniform

51. The heart rates of twelve individuals were measured using two different pieces of equipment. A team would like to compare the measured heart rates of the two pieces of equipment. Which hypothesis test would be most appropriate?

a. Paired t-test

b. Two-sample t-test

c. Two-sample p-test

d. Goodness-of-fit test

52. The length of five components was measured by two inspectors (data shown below). Assuming the data follow a normal distribution, what is the test statistic for the appropriate hypothesis test for this scenario? Use $\alpha = 0.01$.

Component	1	2	3	4	5
Inspector A	10.2	9.8	10.1	10.3	10.4
Inspector B	10.0	9.9	10.3	10.0	10.3

a. 0.450

b. 0.650

c. 2.576

d. 4.604

53. A team at a hospital did a project on medication errors. A random sample of medication errors yielded the following results:

Type of error	Number of errors
Incorrect dose	32
Wrong dose	28
Incorrect form	45
Wrong amount	35
Total	**140**

The team would like to test the hypothesis at the 5% significance level that the medication errors occur with equal probabilities. What is the value of the test statistic for this hypothesis test?

a. 0

b. 1.96

c. 4.514

d. 7.815

54. What distribution does the test statistic used in the goodness-of-fit test follow?

a. *t*

b. chi-square

c. *F*

d. Normal

55. A study was done to compare the elasticity of a polymer mortar during a compression test at three different strain rates. Below is the ANOVA table for the statistical analysis. What is your conclusion at the 5% significance level?

Source	dF	SS	MS	F
Strain rate	2	0.1614	0.0807	1.70
Error	12	0.5688	0.0474	
Total	14	0.7302		

a. Reject H_0; conclude that the three means are the same.

b. Reject H_0; conclude that at least one mean is different than the others.

c. Do not reject H_0; conclude that the three means are the same.

d. Do not reject H_0; conclude that at least one mean is different than the others.

56. Which of the following is an assumption required for use and interpretation of a one-way ANOVA?

a. Inhomogeneity of variance

b. Observations are independent

c. Observations are paired

d. Observations are exponentially distributed

57. A two-way ANOVA hypothesis test was used to test the effects of two factors on a response variable. What conclusions can be made from the interaction plot below?

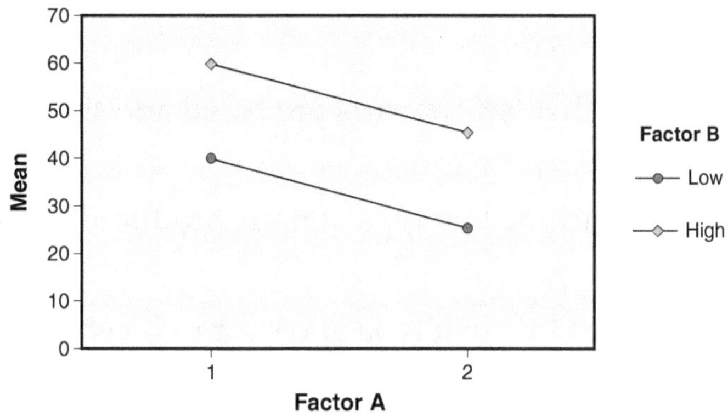

a. High levels of factor B tend to have smaller response values.

b. There is not an interaction effect between factors A and B.

c. There is an interaction effect between factors A and B.

d. Factors A and B do not have an effect on the response variable.

58. Consider the contingency table presented below. A hypothesis test for independence is conducted to determine whether the number of nonconforming parts produced is independent of time of day. What type of distribution does the test statistic for this hypothesis test have?

Time	Conforming parts	Nonconforming parts	Totals
Day	94	5	99
Evening	87	10	97
Overnight	65	23	88
Totals	246	38	284

a. Weibull

b. Normal

c. Binomial

d. Poisson

59. In the previous problem (question 58), the value of the test statistic for the hypothesis test is 19.071. What is the conclusion from this hypothesis test (use $\alpha = 0.05$)?

    a. Reject H_0; conclude that type of parts produced and time of day are independent.

    b. Reject H_0; conclude that type of parts produced and time of day are not independent.

    c. Do not reject H_0; conclude that type of parts produced and time of day are independent.

    d. Do not reject H_0; conclude that type of parts produced and time of day are not independent.

60. The fitted regression equation for two variables x and y is $\hat{y} = 2.5x - 8$. What is the slope of this equation?

    a. –8

    b. 0.89

    c. 1

    d. 2.5

61. A team wants to predict a person's cholesterol based on their weight in pounds. The fitted regression equation is Cholesterol = 140 + 0.23 × Weight. What is the predicted cholesterol of a person who weighs 130 pounds?

    a. 0.23

    b. 140

    c. 162.2

    d. 169.9

62. Paired data collected from a process are: (5.2, 26.7), (6.1, 27.5), (3.2, 24.9), (4.6, 25.5). What is the slope of the linear regression equation for these data?

    a. –0.92

    b. 0.92

    c. 0.96

    d. 21.74

63. A study was done to predict people's cholesterol by their weight. A sample of twenty-one people was collected, and the linear regression equation was found to be Cholesterol = 140 + 0.20 × Weight. The team would like to determine whether there is a significant linear relationship between weight and cholesterol. What is the appropriate alternative hypothesis for this statistical test?

a. $H_a : \beta_1 = 0$

b. $H_a : \beta_0 = 0$

c. $H_a : \beta_1 \neq 0$

d. $H_a : \beta_0 \neq 0$

64. The scatterplot below displays the relationship between two random variables. What is the correlation coefficient of these two variables?

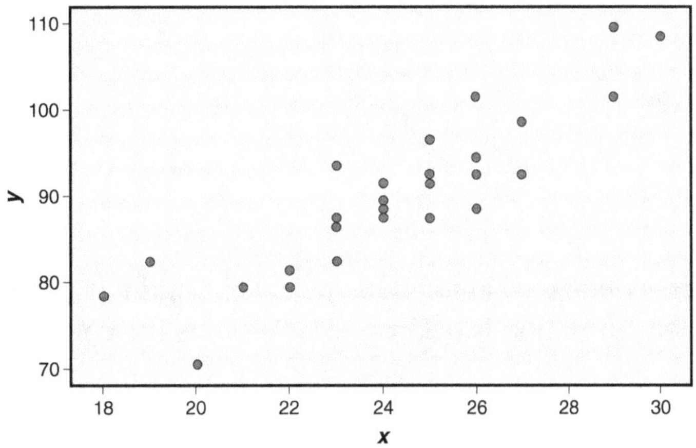

a. –0.88

b. 0

c. 0.88

d. 1.40

65. The scatterplot below displays the relationship between two random variables. What is the correlation coefficient of these two variables?

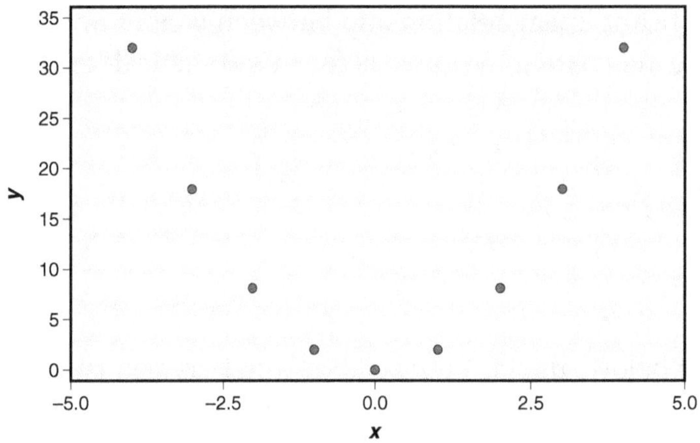

a. −0.98

b. −0.98 and 0.98

c. 0

d. 0.98

66. The fitted regression equation for two variables x and y is $\hat{y} = -5x + 131$. The coefficient of determination is 0.85, or 85%. What is the correlation coefficient?

a. −1.45

b. −0.92

c. 0.85

d. 0.92

67. The correlation coefficient between the age of a car and its cost is $r = -0.85$. This means that:

a. changes in the age of the car cause the cost to change.

b. as the age of a car increases, the cost of the car tends to increase.

c. as the age of a car increases, the cost of the car tends to decrease.

d. the cost of a car is controlled by its age.

68. When the measurement for one sample tends to be dependent on the measure for the previous sample, these data are called:

    a. independent.

    b. autocorrelated.

    c. biased.

    d. inconsistent.

69. Which of the following can help identify special cause variation in a process?

    a. Statistical process control

    b. Regression

    c. Histogram

    d. Capability analysis

70. A control chart was in control, but the operator adjusted the process anyway. What is this called?

    a. Common cause variation

    b. Special cause variation

    c. Overcontrol

    d. Undercontrol

71. What is the principal purpose of control charts?

    a. To identify the source of special cause variation in a process

    b. To adjust a process when it goes out of control

    c. To automate a process

    d. To help process operators recognize the presence of special causes of variation

72. Which of the following events would be considered a source of special cause variation?

    a. Measurement system variability

    b. Incorrect machine settings

    c. Variations in temperature

    d. Slight changes in customer arrivals to a store

73. Which of the following statements about rational subgrouping is correct?

    a. Variability within samples is minimized.

    b. Variability between samples is minimized.

    c. Variability within samples is maximized.

    d. There is no difference between within-sample variability and between-sample variability.

74. Which quality leader introduced the use of control charts to monitor a quality characteristic?

    a. Shewhart

    b. Deming

    c. Juran

    d. Taguchi

75. What probability distribution is required to construct $\bar{X}$ and R control charts?

    a. Binomial

    b. Poisson

    c. Normal

    d. Uniform

76. An $\bar{X}$ and R chart was prepared for an operation using twenty-five samples with seven pieces in each sample. $\bar{X}$ was found to be 24.8 and $\bar{R}$ was 5.50. During production, a $\bar{\bar{X}}$ sample of seven was taken and the pieces measured 25, 32, 35, 28, 27, 24, and 26. At the time this sample was taken,

    a. both the average and range were within the control limits.

    b. neither the average nor range were within the control limits.

    c. only the average was outside the control limits.

    d. only the range was outside the control limits.

77. The diameter of a steel rod is a quality characteristic of interest. Samples of size twelve will be selected in the subgroups. Which of the following control charts is preferred to monitor the process variability?

    a. $\bar{X}$ and R chart

    b. $\bar{X}$ and s chart

c. *p*-chart

d. *c*-chart

78. An $\overline{X}$ and *s* chart was prepared for an operation using twenty-five samples with six pieces in each sample. $\overline{\overline{X}}$ was found to be 12.68 and $\overline{s}$ was 2.45. During production, a sample of six was taken and the pieces measured 11, 13, 14, 14, 10, and 12. At the time this sample was taken:

    a. both the average and standard deviation were within the control limits.

    b. neither the average nor standard deviation were within the control limits.

    c. only the average was outside the control limits.

    d. only the standard deviation was outside the control limits.

79. A company would like to monitor a continuous quality characteristic of a product. However, it is expensive to obtain the measurements of this quality characteristic. Which of the following control charts is most appropriate to monitor this process?

    a. I-MR chart

    b. $\overline{X}$ and *R* chart

    c. $\overline{X}$ and *s* chart

    d. *p*-chart

80. A hospital is monitoring the surgeries that result in surgical complications. The number of surgeries each month and those resulting in a complication were recorded. The number of surgeries each month can vary. Which of the following control charts is most appropriate to monitor this process?

    a. $\overline{X}$ and *R* chart

    b. *p*-chart

    c. *np*-chart

    d. *c*-chart

81. A factory collected data on the number of nonconforming parts and constructed a *p*-chart. Fifteen samples of size 150 were collected. They determined that the average fraction of nonconforming parts was $\overline{p} = 0.037$. During production, a sample of 150 parts was taken, of which eleven were nonconforming. At the time this sample was taken:

    a. the sample was within the control limits.

    b. the sample was outside the control limits.

c. the upper control limit was 0.073.

d. the lower control limit was –0.01.

82. What is the upper control limit for a p-chart when the average daily production is 3575 units with an established fraction defective of 0.048?

a. 0.0600

b. 0.7053

c. 0.0587

d. 0.6893

83. A team at a factory would like to monitor the number of defects in a TV screen. Twenty-five samples of five TV screens are inspected. Which of the following control charts is most appropriate for this scenario?

a. $\bar{X}$ and R chart

b. p-chart

c. np-chart

d. c-chart

84. A team at a factory is monitoring the average number of defects in TV screens. Thirty-five TV screens are randomly selected each day over a 25-day period. The average number of defects per TV screen was found to be $\bar{u} = 0.73$. What is the upper control limit for a corresponding u-chart?

a. 0.22

b. 0.30

c. 1.16

d. 1.24

85. If a process is out of control, the probability that a single point on the $\bar{X}$ chart will fall between plus two sigma and the upper control limit is:

a. 0.1587.

b. 0.2240.

c. 0.3413.

d. unknown.

86. If a process is in control, it is desirable to:

   a. adjust the process when a point is not on target.

   b. have a small average run length.

   c. have a large average run length.

   d. stop monitoring the process.

87. Consider the following MR chart of a normally distributed process. What conclusions can be drawn from this control chart?

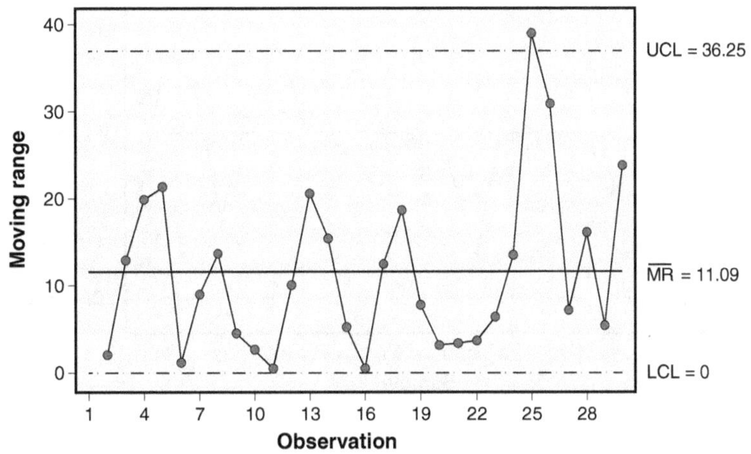

   a. The process is out of control.

   b. The process is stable.

   c. The process is operating within the specifications.

   d. The process is operating outside the specifications.

88. Consider the following *p*-chart. What conclusions can be drawn from this control chart?

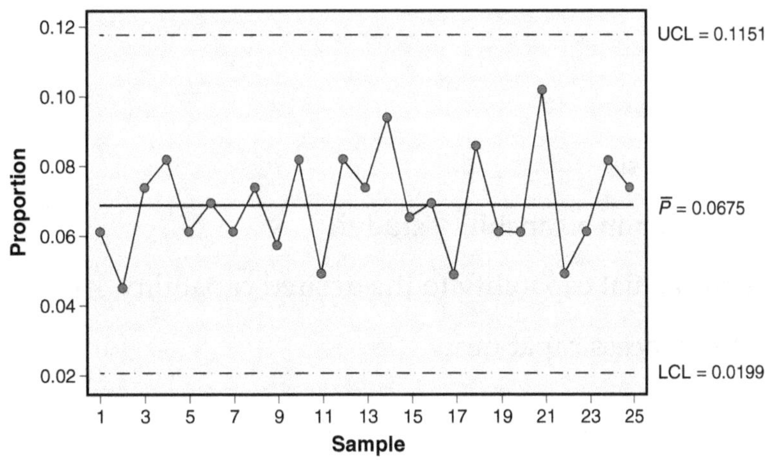

a. The process is out of control.

b. The process is stable.

c. The process is operating within the specifications.

d. The process is operating outside the specifications.

89. Consider the following $\overline{X}$ chart of a normally distributed process. What conclusions can be drawn from this control chart?

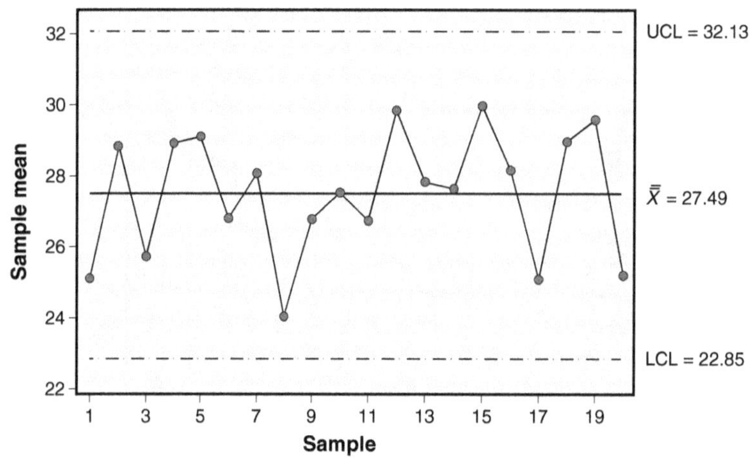

a. The process is out of control.

b. The process is stable.

c. The process is operating within the specifications.

d. The process is operating outside the specifications.

90. A manufacturing company received an order for a build-to-order product from a customer. The company would like to monitor the process; which of the following tools is most appropriate?

a. $\overline{X}$ and s control chart

b. $\overline{X}$ and R control chart

c. Short-run SPC

d. Capability analysis

91. What is the first step in a capability study?

a. Compare the actual capability to the desired capability.

b. Measure the process capability.

c. Estimate the process parameters.

d. Verify that the process is stable.

92. A team would like to determine whether a process is able to meet customer specifications. What quality tool should they use?

a. Flowchart

b. Cause-and-effect diagram

c. Capability analysis

d. Regression analysis

93. The dimension of a component has specifications 5.5 ± 0.25. The process data are normally distributed with mean 5.45 and standard deviation 0.085. What fraction of components will have this dimension outside the specification limits?

a. 0.0002

b. 0.0096

c. 0.9904

d. 0.9998

94. The diameter of a stainless-steel rod is an important quality characteristic. The rod has specifications 2.50 ± 0.05. Data from the process indicate the distribution is normally distributed, and an $\bar{X}$ and R chart indicates the process is stable. The control charts used a sample of size seven and found that $\bar{\bar{X}} = 2.5014$ and $\bar{R} = 0.0568$. What fraction of steel rods are within the specification limits?

a. 0.0071

b. 0.0175

c. 0.9825

d. 0.9896

95. Which of the following is the correct formula to find the specification limits of a quality characteristic?

a. $\mu \pm 3\sigma$

b. $\bar{x} \pm 3\frac{\sigma}{\sqrt{n}}$

c. $\frac{USL - LSL}{6\sigma}$

d. Specification limits are determined externally and do not have a generic formula.

96. A stable, normally distributed process has specifications 21.35 ± 4.50. A sample of data from the process had a mean $\bar{X} = 20.98$ and standard deviation $\sigma = 1.10$. Find C_p and C_{pk}.

a. $C_p = 1.25$; $C_{pk} = 1.36$

b. $C_p = 1.36$; $C_{pk} = 1.25$

c. $C_p = 1.36$; $C_{pk} = 1.48$

d. $C_p = 2.73$; $C_{pk} = 1.36$

97. A stable process has specifications 14 ± 5. The process is normally distributed with mean 13 and standard deviation 1.5. Is this process considered capable?

a. Yes, since $C_p = 0.89$.

b. Yes, since $C_{pk} = 1.11$.

c. No, since $C_{pk} = 0.89$.

d. No, since $C_{pk} = 1.33$.

98. The diameter of a stainless-steel rod is an important quality characteristic. The rod has specifications 2.50 ± 0.05. Data from the process indicate the distribution is normally distributed, and an $\bar{X}$ and R chart indicates the process is stable. The control charts used a sample of size seven and found that $\bar{X} = 2.5014$ and $\bar{R} = 0.0568$. What is the capability of this process?

a. 0.77

b. 0.79

c. 0.82

d. 0.98

99. Which of the following metrics is recommended to evaluate process capability when the process is not in statistical control?

a. C_p

b. P_p

c. C_{pk}

d. C_r

Use the following scenario to answer the following three problems (questions 100–102).

A textile company is designing an experiment to test the effects of temperature, pressure, and time on tensile properties of a fiber composite. Each of these factors will be run at a high and low level. A full-factorial design is used with four replicates per run.

100. How many treatment combinations are in this experiment?

    a. 3

    b. 4

    c. 8

    d. 9

101. How many factors are in this experiment?

    a. 2

    b. 3

    c. 4

    d. 12

102. Variation observed in the readings of the replicates for each treatment combination is referred to as what?

    a. Interaction

    b. Experimental error

    c. Capability

    d. Noise variable

103. What is the first step in designing a successful experiment?

    a. Choose the factors and responses.

    b. Select the experimental design.

    c. Define the method of measurement.

    d. State the objective of the experiment.

104. What technique in experimental design will reduce the effect of uncontrolled variables that are not part of the experiment but might affect the response variable?

    a. Confounding

    b. Blocking

c. Randomization

d. Interaction

105. An experiment is being conducted in a factory. Each experimental run takes several hours, so the experiment will last several days. What design technique would help in this experiment?

a. Confounding

b. Blocking

c. Randomization

d. Replication

106. An experiment is being conducted on the yield of a substance. Two factors are believed to have an effect on the yield. What type of design would be appropriate for this experiment?

a. Fractional factorial design

b. Interaction design

c. Factorial design

d. Confounded design

107. Consider the following results of a 2^3 full-factorial design. Assuming normality and equal variance assumptions are valid, which terms have a significant effect on the response variable?

Source	df	SS	MS	*F*	*p*-value
A	1	35.03	35.03	264.38	0.000
B	1	30.54	30.54	230.49	0.000
C	1	0.10	0.10	0.75	0.400
AB	1	2.98	2.98	22.49	0.001
AC	1	0.02	0.02	0.15	0.693
BC	1	0.03	0.03	0.23	0.646
ABC	1	0.002	0.002	0.15	0.902
Error	8	1.06	0.1325		
Total	15	69.76			

a. A and B only

b. A, B, and AB

c. All main effects

d. C, AC, BC, and ABC

108. Which of the following graphical tools can help assess the assumption of independence when analyzing the results of a full-factorial design?

a. Contour plot

b. Normal probability plot

c. Histogram

d. Run chart

109. A 2^4 full-factorial experiment is conducted with a single replicate. If all main effects and interactions are included in the model, how many degrees of freedom does the error term have?

a. 0

b. 1

c. 8

d. 15

110. An experiment has eight factors at two levels each. The experiment has 128 runs. What is the experimental design called?

a. Full-factorial design

b. Half-fractional factorial design

c. Taguchi design

d. Quarter-fractional factorial design

111. An advantage of using a full-factorial design over a fractional factorial design is that:

a. all main effects and two-factor interactions are estimable.

b. it has projection properties.

c. it requires fewer runs.

d. effects are confounded.

112. In a resolution IV fractional factorial experiment, main effects are confounded with:

 a. three-factor and higher interactions.

 b. other main effects.

 c. two-factor and higher interactions.

 d. no other effects.

113. Consider an experiment with five factors each at two levels. Using the defining relation I = ABD = ACE = BCDE, with which effects is main effect B aliased?

 a. BD and CE

 b. BC, DE, ABE, and ACD

 c. AD and CDE

 d. B is not aliased with other effects.

114. An experiment is being conducted to find the optimal ingredients for a cake mix. The experimenters want the taste of the cake to be unaffected by small variations in oven temperature in the customers' homes. This is a problem of:

 a. interaction.

 b. confounding.

 c. measurement error.

 d. robustness.

115. A team conducted an experiment to determine the optimal amount of three ingredients in a cake to maximize the taste. Two noise factors (oven temperature and baking time) were also included. A crossed-array design was used. The analysis indicated that there was a significant interaction between one of the ingredients and the oven temperature. Which of the following statements is true?

 a. There is no robust design problem for this experiment.

 b. It may be possible to find settings of the ingredients that are robust to the oven temperature.

 c. The oven temperature does not have an effect on the taste of the cake.

 d. The three ingredients are uncontrollable factors in this experiment.

SOLUTIONS

1. a; Discrete data can only take on countable values. The result of the medical test is pass/fail. Errors in the medical record and the number of patient falls have possible values of 0, 1, 2, ... Patient weight can be an infinite number of values and is therefore continuous data. [VI.A.1]

2. c; Length, volume, and time can take on values over an interval of numbers. There are infinitely many values between any two values on their respective scales. Number of nonconforming components, however, can only take on a set of countable values, so it is discrete data. [VI.A.1]

3. a; The color of a car is classified into categories and there is no order implied. [VI.A.2]

4. d; Height has meaningful differences and an absolute zero exists. [VI.A.2]

5. b; Rating of a service is classified into categories where order is important. The precise difference between values is not defined. [VI.A.2]

6. c; An architecture, scope, and framework for the data collection program is first developed that identifies the data organization require for current and future business needs. That is followed by the use of the extract, transform, load (ETL) or extract, load, transform (ELT) process. The Data Transform phase's main goal is preparing data to fit the parameters of another system. [VI.A.3]

7. c; The goodness-of-fit test confirms which distribution best "fits" the data set. That determination can be made based on probability plot data fit and/or the p value for the goodness of fit test. Based on the table values, the Weibull distribution has the highest p value and is the best fit for data set. [VI.A.3]

8. b; Coding data can simplify recording and analysis. Sometimes it is useful to code data using an algebraic transformation. Suppose a set of data has mean μ and standard deviation σ. A new data set may be created using the formula $y = ax + b$. Each element of the new set is formed by multiplying an element of the original set by a, then adding b. [VI.A.3]

9. c;

$$\bar{x} = \frac{5+8+12+3+2+7+6+5}{8} = 6 \text{ minutes.}$$

The median is the value that divides the ordered data into two equal parts. The ordered data are: 2, 3, 5, 5, 6, 7, 8, 12. Since there is an even number of observations, the median is the average of the two central values. The location of the median is

$$\frac{n+1}{2}=\frac{9}{2}=4.5.\ \mathrm{M}=\frac{5+6}{2}=5.5\text{ minutes.}$$

[VI.A.6]

10. d;

$$\bar{x}=\frac{\Sigma_{i=1}^{n}x_i}{n}=\frac{3.2+3.1+3.4+3.1+3.2+3.3+3.2}{7}=3.21486\text{ inches.}$$

$$s=\sqrt{\frac{\Sigma_{i=1}^{n}(x_i-\bar{x})^2}{n-1}}=\sqrt{\frac{(3.2-3.21486)^2+\ldots+(3.2-3.21486)^2}{7-1}}=\sqrt{\frac{0.06857}{6}}=0.1069\text{ inches.}$$

[VI.A.6]

11. c; The population mean and median are not equal, so the distribution cannot be symmetric. The mean is sensitive to the data, while the median is a robust statistic and is not heavily influenced by extreme observations. Therefore, a much smaller mean indicates that the data are skewed to the left. [VI.A.5]

12. b; Under the central limit theorem, the distribution of sample averages tends toward a normal distribution as the sample size increases. Although the exponential distribution is right skewed, the distribution of sample averages is approximately normally distributed for large sample sizes. In this case $n = 40$. [VI.A.6]

13. c; The formula for the population variance is defined as

$$\sigma^2=\frac{\Sigma_{i=1}^{N}(x_i-\mu)^2}{N}.$$

Note that the sample variance is defined as

$$s^2=\frac{\Sigma_{i=1}^{n}(x_i-\bar{x})^2}{n-1}.$$

[VI.A.6]

14. b; Side-by-side box plots can be used to easily compare any differences between the wait times by day of the week. Histograms are more suited for describing the distribution of one population. Scatter plots are used to examine the relationship between two continuous variables. Probability plots are used to determine whether data follow a certain distribution, for example, the normal distribution. [VI.A.7]

15. a; A box plot divides a sorted data set into four regions. The lines in a box plot are marked by the minimum, Q1, median, Q3, and the maximum. The mean is not typically identified on a box plot. [VI.A.7]

16. a; A normal probability plot can help determine whether a set of data comes from a population with a normal distribution. The normal distribution is a reasonable model if the data points fall along a straight line. The data points in this normal probability plot do not fall in a straight line, so the normal distribution would *not* be an appropriate model for the time to failure. [VI.A.7]

17. d; The average temperature of fifteen randomly selected cups of coffee is a *statistic*. A statistic is a characteristic of a sample and is used to estimate the population parameter. In this case, the sample mean temperature of the fifteen cups of coffee can be used to estimate the temperature of all coffee served in the restaurant. [VI.B.1]

18. c; A *parameter* is a characteristic of a population. It is a quantity that describes characteristics of a population. The value of a population parameter is frequently unknown and must be estimated by a sample statistic. [VI.B.1]

19. b; Let N be the event that a nonconforming part is made and let O be the event that the part was made overnight. The given condition is that the part is made overnight, so

$$P(N|O)=\frac{P(N\cap O)}{P(O)}=\frac{23/284}{88/284}=0.26.$$

[VI.B.3]

20. c; P(Not defective) = 1 – P(Defective) = 1 – 0.02 = 0.98. [VI.B.3]

21. d; There is not enough information provided to determine the probability that both events A and B occur. If it were stated that A and B were independent, then the multiplication rule would apply: $P(A \cap B) = P(A) \times P(B)$. [VI.B.3]

22. a; Since events A and B are mutually exclusive, they cannot occur simultaneously. Therefore, $P(A \cap B) = 0$. Note that $P(A \cup B) = P(A) + P(B) = 0.40 + 0.13 = 0.53$, since A and B are mutually exclusive. Recall that probabilities only take on values between 0 and 1. [VI.B.3]

23. b; Let T represent the time between arrivals. T has an exponential distribution with mean $\mu = 5$ minutes. Since

$$\mu=\frac{1}{\lambda},\ \lambda=\frac{1}{\mu}=\frac{1}{5}=.02$$

The cumulative density function for an exponentially distributed random variable T is given by

$$F(x) = P(T \le x) = 1 - e^{-\lambda x}.$$

Therefore,

$$P(T < 3) = 1 - e^{-\lambda \times 3} = 1 - e^{-0.2 \times 3} = 1 - 0.5488 = 0.4512.$$

[VI.C.1]

24. a; X has a normal distribution with mean $\mu = 15$ and standard deviation = 2. The probability of interest is $P(X > 18)$. Transform X into the random variable Z using the relationship

$$Z = \frac{X - \mu}{\sigma}.$$

Then use a standard normal table to find the appropriate probability. From the properties of the cumulative distribution function, $P(X > a) = 1 - P(X < a)$.

$$P(X > 18) = P\left(\frac{X - \mu}{\sigma} > \frac{18 - 15}{2}\right) = P(Z > 1.50)$$
$$= 1 - P(Z < 1.50) = 1 - 0.9332 = 0.0668.$$

[VI.C.1]

25. d; The normal distribution, *t*-distribution, and uniform distribution are all symmetric probability distributions. Asymmetric distributions include the exponential distribution, Weibull distribution, chi-square distribution, and lognormal distribution. [VI.C.1]

26. b; For overfilled, one would only want to examine the proportion of bottles with more than 24.08 oz. Let X represent the amount in a given bottle. Then for $P(X > USL)$.

$$P(X > USL) = P(X > 24.08) = P\left(\frac{X - \mu}{\sigma} > \frac{24.08 - 24.01}{0.025}\right)$$
$$= P(Z > 2.80) = 1 - P(Z < 2.80) = 0.00256.$$

[VI.C.1]

27. d; The *t*-distribution is symmetric, bell-shaped, and has mean $\mu = 0$, like the standard normal distribution. However, the tails of the *t*-distribution are longer or heavier compared to the standard normal distribution. [VI.C.1]

28. d; Let X be the number of nonconforming units in a day. X has a Poisson distribution with rate $\lambda = 5$.

$$\begin{aligned}P(X \geq 2) &= 1 - P(X \leq 1) = 1 - P(X = 0) - P(X = 1)\\ &= 1 - \frac{e^{-5}5^0}{0!} - \frac{e^{-5}5^1}{1!}\\ &= 1 - 0.0067 - 0.0337 = 0.9596.\end{aligned}$$

[VI.C.2]

29. a; The hypergeometric distribution is appropriate when the sample is taken without replacement from a finite population. The binomial distribution is appropriate when sampling is done with replacement. The normal distribution is not appropriate, since it is a continuous probability distribution. [VI.C.2]

30. b; Let X represent the number of invoices with an error. In this scenario, X has a binomial distribution with $n = 15$ and $p = 0.10$. The probability of interest is $P(X = 4)$.

$${}_nC_x = \binom{15}{4} = \frac{15!}{4!(15-4)!} = 1365.$$

$$\begin{aligned}P(X = 4) &= {}_nC_x p^x (1-p)^{n-x} = {}_{15}C_4 (0.10)^4 (1-0.10)^{15-4}\\ &= 1365(0.0001)(0.3138) = 0.0428.\end{aligned}$$

[VI.C.2]

31. b;

$$E(X) = \mu_X = \Sigma x f(x) = 0(0.86) + 1(0.06) + 2(0.04) + 3(0.03) + 4(0.01) = 0.27.$$

[VI.C.2]

32. d; The central limit theorem states that if the sample size n is sufficiently large, then the sample mean $\bar{X}$ follows approximately a normal distribution with mean $\mu_{\bar{X}} = \mu$ and standard deviation $\sigma_{\bar{X}} = \sigma/\sqrt{n}$. In this case, $\mu = 24$, $\sigma = 5.5$, and $n = 32$. Therefore, $\mu_{\bar{X}} = \mu = 24$. [VI.C.2]

33. b; Since the underlying distribution is normal, then the sampling distribution of $\bar{X}$ also follows a normal distribution with mean $\mu_{\bar{X}} = \mu$ and standard deviation $\sigma_{\bar{X}} = \sigma/\sqrt{n}$. Note that the large sample size requirement is not necessary here, since the filling process has a normal distribution. In this population, $\mu = 24$, $\sigma = 0.05$, and $n = 16$. Therefore, $\mu_{\bar{X}} = \mu = 24$ and $\sigma_{\bar{X}} = \sigma/\sqrt{n} = 0.05/\sqrt{16} = 0.0125$.

The probability of interest is

$$P(\bar{X} > 24.04) = P\left(\frac{\bar{X} - \mu_{\bar{X}}}{\sigma_{\bar{X}}} > \frac{24.04 - 24}{0.05/\sqrt{16}}\right) = P(Z > 3.20)$$

$$= 1 - P(Z < 3.20) = 1 - 0.9993 = 0.0007.$$

[VI.C.2]

34. a; In this problem $\mu = 58$, $\sigma^2 = 64$, and $n = 16$. The standard error is defined as

$$\text{s.e.}(\bar{X}) = \sigma/\sqrt{n} = \sqrt{64}/\sqrt{16} = 8/4 = 2.$$

[VI.D.1]

35. d; Recall that the true value of the population proportion is unknown. Answer (d) describes the practical interpretation of the confidence interval: with 95% confidence, the population proportion will lie between 0.02 and 0.12. [VI.D.1]

36. a; 11.73 mm is a point estimate for the average length of the component. Recall that a parameter is a characteristic of a population. The sample mean is often a point estimator for the population mean. [VI.D.1]

37. b; The population variance is unknown, so the sample standard deviation is used to estimate the population standard deviation. The $100(1 - \alpha)\%$ confidence interval for μ then is

$$\bar{x} - t_{\alpha/2,n-1}\frac{s}{\sqrt{n}} \le \mu \le \bar{x} + t_{\alpha/2,n-1}\frac{s}{\sqrt{n}}$$

$$\bar{x} = 4.3,\ s = 0.51,\ n = 15,\ \alpha/2 = 0.005,\ t_{0.005,14} = 2.977$$

$$4.3 - 2.977\left(\frac{0.51}{\sqrt{15}}\right) \le \mu \le 4.3 + 2.977\left(\frac{0.51}{\sqrt{15}}\right)$$

$$3.908 \le \mu \le 4.692.$$

There is 99% confidence that the mean tensile strength is between 3.908 and 4.692 MPa. [VI.D.1]

38. d; As stated in the problem, the estimate should be within 0.15 MPa of the true average. This means the estimate can be at most 0.15 MPa less than the average, or at most 0.15 MPa greater than the true average. So, the margin of error is $E = 0.15$. Furthermore, $\sigma = 0.50$, $\alpha = 0.05$, $Z_{\alpha/2} = Z_{.025} = 1.96$. The minimum sample size needed is

$$n \geq \left(\frac{\sigma Z_{\alpha/2}}{E}\right)^2 = \left(\frac{0.50(1.96)}{0.15}\right)^2 = 42.68.$$

Therefore, the minimum sample size of cement would be $n = 43$. [VI.D.1]

39. d;

$$\bar{x} = \frac{\Sigma_{i=1}^{n} x_i}{n} = \frac{10.37+11.50+9.80+10.65+11.95+10.15+9.52}{7} = 10.563 \text{ V}.$$

$$s^2 = \frac{\Sigma_{i=1}^{n}(x_i - \bar{x})^2}{n-1} = \frac{(10.37-10.563)^2 + \ldots + (9.52-10.563)^2}{7-1} = \frac{4.687}{6} = 0.781 \text{ V}^2.$$

The 100(1 – α)% two-sided confidence interval on the population variance σ^2 is given by

$$\frac{(n-1)s^2}{\chi^2_{\alpha/2,k}} < \sigma^2 < \frac{(n-1)s^2}{\chi^2_{1-\alpha/2,k}}$$

$$n=7,\ s^2=0.781,\ \alpha=0.05,\ k=n-1=6,\ \chi^2_{0.025,6}=14.449,\ \chi^2_{0.975,6}=1.237.$$

$$\frac{(7-1)0.781}{14.449} < \sigma^2 < \frac{(7-1)0.781}{1.237}$$

$$0.324 < \sigma^2 < 3.788.$$

There is 95% confidence the true population variance of the voltage of the power supply lies within 0.324 V^2 and 3.788 V^2. [VI.D.1]

40. a; A one-sample *t*-test is appropriate where $H_0 : \mu = 18$ and $H_a : \mu \neq 18$. The test statistic for this hypothesis test is

$$t_0 = \frac{\bar{x} - \mu_0}{s/\sqrt{n}} = \frac{17.98-18}{0.03/\sqrt{24}} = -3.266.$$

Note: H_0 is rejected if $t_0 < -t_{\alpha/2,n-1}$ or $t_0 > t_{\alpha/2,n-1} = t_{0.025,23} = 2.069$.

Since $-3.266 < -2.069$, H_0 is rejected, and the conclusion is that the mean volume of soda is not equal to 18 oz. [VI.D.2]

41. b; The *p*-value for the hypothesis test is 0.003, which is less than the significance level 0.05. Therefore, there is sufficient evidence to reject the null hypothesis and conclude that the mean volume is different than 18 oz. A 95% confidence interval of the population mean, therefore, would *not* contain 18. [VI.D.2]

42. c; The appropriate hypothesis test is a *t*-test, since it is a test for a single population mean μ and the population variance is unknown. A left-tailed test is appropriate, since the team would like to assess whether the mean is 25 mph or less. Note: The hypotheses for this test are $H_0 : \mu = 25$ versus $H_a : \mu < 25$. The test statistic is

$$t_0 = \frac{\bar{x} - \mu}{s / \sqrt{n}} = \frac{24.50 - 25}{2.3 / \sqrt{20}} = -0.97.$$

Using a significance level of 5%, H_0 is rejected if $t_0 < t_{\alpha,n-1} = -t_{0.05,19} = -1.729$. Since $-0.97 > -1.729$, H_0 is not rejected. There is not sufficient evidence to conclude that the mean speed is less than 25 mph. [VI.D.2]

43. c; Let *p* be the true proportion of patient falls in the hospital. x = The number of patient falls = 7. n = Sample size = 56. The sample proportion is then

$$\hat{p} = \frac{x}{n} = \frac{7}{56} = 0.125.$$

The null and alternative hypotheses are $H_0 : p = 0.10$ versus $H_a : p > 0.10$. The significance level is $\alpha = 0.05$. This is a right-tailed test, so the null hypothesis is rejected if the test statistic is greater than $z_\alpha = z_{0.05} = 1.645$. The test statistic is

$$z_0 = \frac{\hat{p} - p_0}{\sqrt{\frac{p_0(1 - p_0)}{n}}} = \frac{0.125 - 0.10}{\sqrt{\frac{0.10(1 - 0.10)}{56}}} = 0.624.$$

Since $0.624 < 1.645$, there is not enough evidence to reject H_0. At the 5% significance level, there is not enough evidence to conclude that the proportion of patient falls is more than 0.10. [VI.D.2]

44. a; The parameter of interest is the difference in average fill volume, $\mu_1 - \mu_2$. The null and alternative hypotheses for the corresponding hypothesis test are $H_0 : \mu_1 - \mu_2 = 0$ versus $H_a : \mu_1 - \mu_2 \neq 0$. Note that the hypothesized value $\Delta_0 = 0$ is not contained in the confidence interval. Therefore, there is evidence that there is a difference in the two population means. [VI.D.2]

45. b; The null and alternative hypotheses for this hypothesis test are $H_0 : \mu_{\text{Before}} - \mu_{\text{After}} = 0$ versus $H_a : \mu_{\text{Before}} - \mu_{\text{After}} > 0$. The significance level is $\alpha = 0.01$. This is a right-tailed test, so the null hypothesis will be rejected if the test statistic is greater than $t_{\alpha,k} = t_{0.01,n_1+n_2-2} = t_{0.01,51} = 2.402$. Note that the test statistic is

$$t_0 = \frac{\bar{x}_{\text{Before}} - \bar{x}_{\text{After}} - 0}{\sqrt{\frac{s^2_{\text{Before}}}{n_{\text{Before}}} + \frac{s^2_{\text{After}}}{n_{\text{After}}}}} = \frac{3.52 - 2.10 - 0}{\sqrt{\frac{2.34}{28} + \frac{2.04}{25}}} = 3.494.$$

Since 3.494 > 2.402, H_0 is rejected, and there is a significant difference in the packaging time. [VI.D.2]

46. b; Let σ_1^2 represent the population variance of the packaging time before the improvement project, and let σ_2^2 represent the population variance of the packaging time after the improvement project. The null and alternative hypotheses are

$$H_0: \sigma_1^2 = \sigma_2^2 \text{ versus } H_a: \sigma_1^2 \neq \sigma_2^2.$$

The test statistic is

$$F_0 = \frac{s_1^2}{s_2^2} = \frac{2.34}{2.04} = 1.147.$$

Note that this is a two-tailed test with significance level $\alpha = 0.10$. The degrees of freedom are $k_1 = n_1 - 1 = 27$ and $k_2 = n_2 - 1 = 24$. The critical values are:

$$F_{\alpha/2,k_1,k_2} = F_{0.05,27,24} = 1.93$$

and

$$F_{1-\alpha/2,k_1,k_2} = F_{0.95,27,24} = \frac{1}{F_{\alpha/2,k_2,k_1}} = \frac{1}{F_{0.05,24,27}} = \frac{1}{1.96} = 0.51.$$

H_0 will be rejected.
If

$$F_0 < F_{1-\alpha/2,k_1,k_2} \text{ or } F_0 > F_{\alpha/2,k_1,k_2}.$$

Since 1.147 > 0.51 and 1.147 < 1.93, H_0 is not rejected, and the conclusion is that there is no significant change in the variance of packaging time before and after the performance improvement project. [VI.D.2]

47. b; Let p_1 represent the proportion of nonconforming units produced by machine 1 and p_2 represent the proportion of nonconforming units produced by machine 2.

$$x_1 = 28,\ n_1 = 1250,\ \hat{p}_1 = \frac{x_1}{n_1} = \frac{28}{1250} = 0.0224.$$

$$x_2 = 18,\ n_2 = 1175,\ \hat{p}_2 = \frac{x_2}{n_2} = \frac{18}{1175} = 0.015319.$$

$\hat{p}$ is the estimate of the overall proportion for the parameter $p_1 - p_2$.

$$\hat{p} = \frac{x_1 + x_2}{n_1 + n_2} = \frac{28+18}{1250+1175} = \frac{46}{2425} = 0.01897.$$

The level of significance is $\alpha = 0.10$. $z_{\alpha/2} = z_{0.05} = 1.645$.

A 100(1 – α)% confidence interval on the parameter $p_1 - p_2$ is

$$(\hat{p}_1 - \hat{p}_2) - z_{\alpha/2}\sqrt{\hat{p}(1-\hat{p})\left(\frac{1}{n_1}+\frac{1}{n_2}\right)} \le p_1 - p_2 \le (\hat{p}_1 - \hat{p}_2) + z_{\alpha/2}\sqrt{\hat{p}(1-\hat{p})\left(\frac{1}{n_1}+\frac{1}{n_2}\right)}$$

$$(0.0224 - 0.015319) - 1.645\sqrt{0.01897(1-0.01897)\left(\frac{1}{1250}+\frac{1}{1175}\right)} \le p_1 - p_2 \le$$

$$(0.0224 - 0.015319) + 1.645\sqrt{0.01897(1-0.01897)\left(\frac{1}{1250}+\frac{1}{1175}\right)}$$

$$-0.002 \le p_1 - p_2 \le 0.016.$$

There is 90% confidence that the difference between the proportion of nonconforming parts is between –0.002 and 0.016. Because 0 is contained in this confidence interval, there is not sufficient evidence to conclude that the two machines have a different proportion of nonconforming parts. [VI.D.2]

48. b; β = P(Type II error) = 1 – Power = 1 – 0.80 = 0.20. The significance level is α = P(Type I error). [VI.D.2]

49. c; A type II error occurs when H_0 is not rejected when in fact H_0 is false. A type I error occurs when H_0 is rejected when, in fact, H_0 is true. [VI.D.2]

50. a; A two-sample *p*-test is used to compare the proportions of defective widgets from two machines. The test statistic for this test is

$$z_0 = \frac{\hat{p}_1 - \hat{p}_2}{\sqrt{\hat{p}(1-\hat{p})\left(\frac{1}{n_1}+\frac{1}{n_2}\right)}},$$

where

$$\hat{p} = \frac{x_1 + x_2}{n_1 + n_2}.$$

This test statistic follows a standard normal distribution for sufficiently large sample sizes. [VI.D.2]

51. a; A paired t-test is most appropriate because the same individuals were measured twice. There is a strong relationship between the two measurements. [VI.D.3]

52. b; A paired t-test is required for this hypothesis test because there is a relationship between two observations. The same component was measured twice, once by each inspector. Therefore, a hypothesis test can be performed on the difference between the length measured by inspector A and the length measured by inspector B.

Component	1	2	3	4	5
Inspector A	10.2	9.8	10.1	10.3	10.4
Inspector B	10.0	9.9	10.3	10.0	10.3
$d_i = A_i - B_i$	0.2	–0.1	–0.2	0.3	0.1

Let μ_1 represent the mean length measured by inspector A and let μ_2 represent the mean length measured by inspector B. Let μ_D represent the true mean difference between the two populations. $\mu_D = \mu_1 - \mu_2$.

The necessary summary statistics are

$$\bar{d} = \frac{\Sigma_{i=1}^{n} d_i}{n} = \frac{0.2+(-0.1)+(-0.2)+(0.3)+(0.1)}{5} = 0.06$$

and

$$s_d = \sqrt{\frac{\Sigma_{i=1}^{n}\left(d_i - \bar{d}\right)^2}{n-1}} = \sqrt{\frac{(0.2-0.06)^2+\ldots+(0.1-0.06)^2}{5-1}} = 0.2074.$$

The null and alternative hypotheses are $H_0 : \mu_D = 0$ versus $H_a : \mu_D \neq 0$. The significance level is = 0.01. This is a two-tailed test, so H_0 will be rejected if the test statistic is less than $-t_{\alpha/2,k}$ or greater than $t_{\alpha/2,k} = -t_{0.005,n-1} = t_{0.005,4} = 4.604$.

$$t_0 = \frac{\bar{d}}{s_d / \sqrt{n}} = \frac{0.06}{0.2074/\sqrt{5}} = 0.65.$$

Since $0.65 < 4.604$, H_0 is not rejected, and it is concluded that there is not a significant difference between the mean length measured by the two inspectors. [VI.D.3]

53. c; A goodness-of-fit test is appropriate here. The team would like to test whether the medication errors occur with equal probabilities. Therefore, the null and alternative hypotheses are $H_0 : p_1 = p_2 = p_3 = p_4 = 0.25$ versus $H_a : p_i \neq 0.25$ for at least one $i = 1, 2, 3, 4$.

The test statistic is

$$\chi_0^2 = \Sigma_{i=1}^{k} \frac{(O_i - E_i)^2}{E_i},$$

where $E_i = (140)(0.25) = 35$ for $i = 1, 2, 3, 4$. Therefore,

$$\chi_0^2 = \frac{(32-35)^2}{35} + \frac{(28-35)^2}{35} + \frac{(45-35)^2}{35} + \frac{(35-35)^2}{35} = 4.514.$$

Note that for a significance level $\alpha = 0.05$, H_0 is rejected if

$$\chi_0^2 > \chi_{\alpha,k-1}^2 = \chi_{0.05,3}^2 = 7.81.$$

Since $4.514 < 7.81$, H_0 is not rejected, and it is concluded that the medication errors occur with equal probabilities. [VI.D.4]

54. b; The test statistic for a goodness-of-fit test is defined as

$$\chi_0^2 = \Sigma_{i=1}^{k} \frac{(O_i - E_i)^2}{E_i}.$$

The test statistic follows a chi-square distribution with $k - 1$ degrees of freedom, where k is the number of categories. [VI.D.4]

55. c; From the ANOVA table and using an F table, one can find the critical value and compare it to the test statistic $F_0 = 1.70$. From the ANOVA table, the degrees of freedom for the critical value are 2 and 12. From an F table, $F_{0.05,2,12} = 19.41$. Since $F_0 < 19.41$, H_0 is not rejected, and it is concluded that there is not a significant difference between the three means. [VI.D.5]

56. b; The assumptions for a one-way ANOVA are that the observations follow a normal distribution, the observations are independent, and the treatments have constant variance. Paired data is not an assumption for ANOVA. [VI.D.5]

57. b; An interaction plot can show you differences in the mean response for the different levels of factors A and B. This interaction plot indicates that there is no significant interaction between factors A and B. As you move across the levels of factor A, the levels of factor B maintain identical patterns. Note that high levels of factor B tend to have higher values of the response, not smaller levels of the response. [VI.D.5]

58. a; The test statistic for a test of independence is

$$\chi_0^2 = \Sigma_{i=1}^{r}\Sigma_{j=1}^{c}\frac{\left(O_{ij} - E_{ij}\right)^2}{E_{ij}}.$$

This test statistic follows a chi-square distribution with $(r - 1)(c - 1)$ degrees of freedom. [VI.D.6]

59. b; The test statistic for the test of independence follows a chi-square distribution with $(r - 1)(c - 1)$ degrees of freedom, where r represents the number of levels of the first factor and c represents the number of levels of the second factor. H_0 will be rejected if

$$\chi_0^2 > \chi^2_{\alpha,(r-1)(c-1)} = \chi_{0.05,2} = 5.991.$$

Since

$$\chi_0^2 = 19.071 > 5.991,$$

H_0 is rejected, and it is concluded that the type of part produced (conforming or nonconforming) and time of day are not independent. [VI.D.6]

60. d; The slope represents the change in the response y for every one unit change in the variable x. Therefore, 2.5 is the slope of the fitted regression equation. [VI.E.1]

61. d; By plugging in the given value of Weight = 130 into the regression equation, the cholesterol can be predicted. Cholesterol = 140 + 0.23 × (130) = 169.9. [VI.E.1]

62. b;

	x	y	xy	x^2	y^2
	5.2	26.7	138.84	27.04	712.89
	6.1	27.5	167.75	37.21	756.25
	3.2	24.9	79.68	10.24	620.01
	4.6	25.5	117.3	21.16	650.25
Σ	19.1	104.6	503.57	95.65	2739.40
Mean	4.775	26.15			

$$S_{xx} = \sum_{i=1}^{4} x_i^2 - \frac{\left(\Sigma_{i=1}^{4} x_i\right)^2}{n} = 95.65 - \frac{(19.1)^2}{4} = 4.4475$$

$$S_{xy} = \sum_{i=1}^{4} x_i y_i - \frac{\left(\Sigma_{i=1}^{4} x_i\right)\left(\Sigma_{i=1}^{4} y_i\right)}{n} = 503.57 - \frac{(19.1)(104.6)}{4} = 4.105$$

The slope of the regression equation can be found using the formula

$$b_1 = \frac{S_{xy}}{S_{xx}} = \frac{4.105}{4.4475} = 0.923.$$

Note that

$$b_0 = \bar{y} - b_1\bar{x} = 26.15 - (0.923)(4.775) = 21.74.$$

Therefore, the full regression equation is

$$\hat{y} = 21.74 + 0.923x.$$

[VI.E.1]

63. c; The appropriate test is a hypothesis test on the slope of the linear regression equation, not the intercept of the linear regression equation. The slope is represented by β_1, and the intercept is represented by β_0. The null hypothesis is $H_0 : \beta_1 = 0$. The alternative hypothesis is $H_a : \beta_1 \neq 0$. [VI.E.1]

64. c; The sample correlation coefficient measures the strength of the linear relationship between two variables. The scatterplot shows that as x increases, y increases. This indicates that there is a positive correlation. The correlation coefficient only takes values between –1 and +1. The closer the value is to –1 or +1, the stronger the linear relationship, so there is a strong positive correlation between these two variables. [VI.E.2]

65. c; The sample correlation coefficient measures the strength of the *linear* relationship between two variables. The scatterplot indicates that there is a quadratic relationship between the two variables, but this relationship is not linear. Therefore, the correlation coefficient r is 0. [VI.E.2]

66. b; The correlation coefficient equation is

$$r = \pm\sqrt{R^2}.$$

The coefficient of determination is $R^2 = 0.85$. The slope of the regression equation is –5, indicating that there is a negative linear relationship between x and y.

Therefore, the correlation coefficient is

$$r = -\sqrt{0.85} = -0.92.$$

[VI.E.2]

67. c; Since the correlation coefficient is negative, there is an inverse linear association between the age and cost of the car. As the age of the car increases, the cost tends to decrease. Note that although there is a linear correlation, this does not imply that there is a causal relationship. An increase in the age of the car does not cause the cost to decrease. [VI.E.2]

68. b; Data that are autocorrelated means that the data in the previous sample affect the data in the next sample. [VI.E.3]

69. a; Statistical process control is a statistical technique that can help identify special cause variation. Regression analysis describes the relationship between one or more predictor variables and a response variable. Histograms are used to visualize data. Capability analysis is used to determine whether a process is able to meet certain requirements. [VI.F.1]

70. c; Overcontrol, also called *overadjustment*, occurs when a process operator tries to adjust a process in response to common cause variation. In this scenario, the control chart has not indicated out-of-control behavior, but the operator adjusted the process anyway. [VI.F.2]

71. d; Control charts indicate whether a process is out of control, but they cannot identify what caused the out-of-control behavior. Control charts do not automatically adjust the process. It is necessary for the process operator to determine and correct out-of-control behavior when it is indicated by a control chart. [VI.F.2]

72. b; Common cause variability is variation that is inherent in the process and not controllable by operators. Variation in ambient temperature, the measurement system, and customer arrivals are natural variability in a process. An incorrect setting on a machine, however, is an unusual event that can be adjusted by the operator. An incorrect setting is not inherent to the process and thus is a source of special cause variation. [VI.F.2]

73. a; Rational subgrouping involves selecting samples such that if there is special cause variation present in the process, there will be a greater probability of variation between samples, while the variation within samples is small. Therefore, the variability within samples is minimized and the variability between samples is maximized. [VI.F.4]

74. a; Walter A. Shewhart introduced the concept of control charts in the 1920s. [VI.F.5]

75. c; $\bar{X}$, R, s, I, and MR control charts require that the data under observation follow a normal distribution. Violations of this assumption will lead to poor performance of the control charts, particularly with I-MR charts. [VI.F.5]

76. b; The summary statistics for the sample are

$$\bar{x} = \frac{25+32+35+28+27+24+26}{7} = 28.14; \; R = 35-24 = 11.$$

From the information given,

$$\bar{\bar{x}} = 24.8, \; \bar{R} = 5.50, \; n = 7.$$

The control limits for the $\bar{X}$ chart are

$$\text{UCL} = \bar{\bar{x}} + A_2\bar{R} = 24.8 + 0.419(5.50) = 27.10$$

$$\text{LCL} = \bar{\bar{x}} - A_2\bar{R} = 24.8 - 0.419(5.50) = 22.50$$

The control limits for the R chart are

$$\text{UCL} = D_4\bar{R} = (1.924)(5.50) = 10.58$$

$$\text{LCL} = D_3\bar{R} = (0.076)(5.50) = 0.418$$

A_2, D_3, and D_4 can be found in Appendix B of *The ASQ Certified Quality Engineer Handbook*, Fifth Edition for $n = 7$.

Therefore, since $\bar{X} = 28.14$ is greater than the UCL for the $\bar{X}$ chart, and $R = 11$ is greater than the UCL for the R chart, neither the average nor the range were within the control limits. [VI.F.5]

77. b; The diameter of the steel rod is variables data. Therefore, the p-chart and c-chart are not appropriate to monitor process variability. The sample size is relatively large ($n = 12$); therefore, the sample standard deviation is a better estimate of the process standard deviation compared to the range. In general, the sample standard deviation should be used instead of the range in estimating process variability. [VI.F.5]

78. a; The summary statistics for the sample are

$$\bar{x} = \frac{11+13+14+14+10+12}{6} = 12.33; \; s = \sqrt{\frac{(11-12.33)^2 + \ldots + (12-12.33)^2}{6-1}} = 1.633.$$

From the information given,

$$\bar{\bar{x}} = 12.68, \ \bar{s} = 2.45, \ n = 6$$

The control limits for the $\bar{X}$ chart are

$$\text{UCL} = \bar{\bar{x}} + A_3\bar{s} = 12.68 + 1.287(2.45) = 15.83$$

$$\text{LCL} = \bar{\bar{x}} - A_3\bar{s} = 12.68 - 1.287(2.45) = 9.53$$

The control limits for the s-chart are

$$\text{UCL} = B_4\bar{s} = (1.970)(2.45) = 4.83$$

$$\text{LCL} = B_3\bar{s} = (0.03)(2.45) = 0.07$$

A_3, B_3, and B_4 can be found in Appendix B of *The ASQ Certified Quality Engineer Handbook*, Fifth Edition for $n = 6$.

Since $\bar{X} = 12.33$ is within the UCL and LCL for the $\bar{X}$ chart, and $s = 1.633$ is within the UCL and LCL for the s-chart, both the average and standard deviation were within the control limits. [VI.F.5]

79. a; Because it is expensive to obtain the measurement of the quality characteristic, the subgroup should consist of a single observation ($n = 1$). Therefore, I-MR control charts are appropriate. Note that the quality characteristic is continuous; thus the p-chart is not appropriate. [VI.F.5]

80. b; The quality characteristic of interest is whether or not a surgery had complications. This is attribute data, so an $\bar{X}$ and R chart is not appropriate. A c-chart is also not appropriate, as that control chart monitors nonconformities. Because the number of surgeries can vary each month, the p-chart is preferred over the np-chart. While np-charts are often easier to interpret, they should be used when the sample sizes are the same to avoid misleading results. [VI.F.5]

81. a; From the information provided, $\bar{p} = 0.037$, $m = 15$, $n = 150$. The new sample fraction nonconforming is

$$\hat{p} = \frac{x}{n} = \frac{11}{150} = 0.073.$$

Part VI Solutions

The control limits for the *p*-chart are

$$UCL = \bar{p} + 3\sqrt{\frac{\bar{p}(1-\bar{p})}{n}} = 0.037 + 3\sqrt{\frac{(0.037)(1-0.037)}{150}} = 0.083$$

$$LCL = \bar{p} - 3\sqrt{\frac{\bar{p}(1-\bar{p})}{n}} = 0.037 - 3\sqrt{\frac{(0.037)(1-0.037)}{150}} = -0.01 \rightarrow 0.$$

Since $\hat{p} = 0.073$ lies within the UCL and LCL of the *p*-chart, the sample is within the control limits. Note that the formula for the LCL leads to a negative number. By convention, this value is changed to 0, since you cannot have a negative proportion. [VI.F.5]

82. c; From the information provided, $\hat{p} = 0.048$, and $n = 3575$.

$$UCL = \bar{p} + 3\sqrt{\frac{\bar{p}(1-\bar{p})}{n}} = 0.048 + 3\sqrt{\frac{(0.048)(1-0.048)}{3575}} = 0.0587.$$

[VI.F.5]

83. d; Defects on TVs are nonconformities. This is attribute data, so the $\bar{X}$ and *R* charts are not appropriate. A control chart for nonconformities, the *c*-chart, is appropriate here. Note that the *p*- and *np*-charts are for fraction nonconforming. [VI.F.5]

84. c; From the information provided,

$$\bar{u} = 0.73, \; n = 35, \; m = 25$$

$$UCL = \bar{u} + 3\sqrt{\frac{\bar{u}}{n}} = 0.73 + 3\sqrt{\frac{0.73}{35}} = 1.16.$$

[VI.F.5]

85. d; Since the process is unstable (out of statistical control), the actual probability that a point lies within this region is unknown. [VI.F.6]

86. c; The average run length is the number of cycles, time periods, or samples that elapse before the process signals out of control. When the process is in control, it is desired that the average run length be large. If the process is in control, operators should not adjust the process if a point is not right on target. This overadjustment can destabilize a stable process. [VI.F.6]

87. a; A point is above the upper control limit, indicating that the process is out of statistical control. Note that a control chart does not provide information on whether the process is operating within the specification limits. A process

could be out of statistical control, but still be operating within the specification limits. Similarly, a process could be in control, but not be within the specification limits. [VI.F.6]

88. a; Beginning with the fourth data point, there is a sequence of sixteen points alternating up and down. One of the sensitizing rules to detect out-of-control behavior is that there are fourteen points in a row alternating up and down. Note that a control chart does not provide information on whether the process is operating within the specification limits. A process could be out of statistical control, but still operating within the specification limits. Similarly, a process could be in control, but not be within the specification limits. [VI.F.6]

89. b; There is no unusual behavior indicated in the control chart; therefore, the process is in control (stable). Note that a control chart does not provide information on whether the process is operating within the specification limits. A process could be out of statistical control, but still be operating within the specification limits. Similarly, a process could be in control, but not be within the specification limits. [VI.F.6]

90. c; The factory has received a built-to-order product. This is not a process that has a long, continuous production run. Therefore, the classical control charts (for example, $\bar{X}$ and R) are not appropriate for monitoring the process variability. Short-run statistical process control is necessary. Standardized control charts are typically used for short-run processes. [VI.F.7]

91. d; The first step in a capability study is to verify that the process is stable. A stable process is one that does not have any special causes of variation present. If the process is not stable, the capability measures will not be valid estimates of the process capability. [VI.G.1]

92. c; Capability analysis is used to determine whether a process is capable of meeting requirements. The capability of a process can also be evaluated through use of histograms, probability plots, and stem-and-leaf plots. [VI.G.1]

93. b; Let X represent the dimension of the component. The upper specification limit is 5.75 and the lower specification limit is 5.25. First, consider the proportion of components that are within the specification limits. This can be written as $P(5.25 < X < 5.75)$. Since the dimension is normally distributed, the standard normal distribution can be used to determine this fraction:

$$P(LSL < X < USL) = P(X < 5.25 < 5.75) = P\left(\frac{5.25-5.45}{0.085} < \frac{X-\bar{x}}{\hat{\sigma}} < \frac{5.75-5.45}{0.085}\right)$$

$$= P(-2.35 < Z < 3.53) = P(Z < 3.53) - P(Z < -2.53)$$

$$= 0.9998 - 0.0094 = 0.9904.$$

Therefore, the fraction that does not meet the specifications is 1 – 0.9904 = 0.0096.

Approximately 0.96% of components will fall outside specifications for this dimension. [VI.G.2]

94. c; Let X represent the diameter of the steel rod. The upper specification limit is 2.55, and the lower specification limit is 2.45. The proportion of steel rods that are within the specification limits can be written as P(2.45 < X < 2.55). Since the diameter of the steel rod is normally distributed and the process is stable, the standard normal distribution can be used to determine this fraction. From the control charts, an estimate for μ is $\hat{\mu} = \bar{\bar{x}} = 2.5014$. The point estimate for the process standard deviation is

$$\hat{\sigma} = \frac{\bar{R}}{d_2} = \frac{0.0568}{2.704} = 0.021.$$

The value for d_2 can be found in Appendix B of *The ASQ Certified Quality Engineer Handbook*, Fifth Edition, for $n = 7$.

Therefore, the fraction of steel rods that meet the specification limits is

$$\mathrm{P(LSL < X < USL) = P(2.45 < X < 2.55) = P}\left(\frac{2.45 - 2.5014}{0.021} < \frac{X - \bar{\bar{x}}}{\hat{\sigma}} < \frac{2.55 - 2.5014}{0.021}\right)$$

$$= \mathrm{P}(-2.45 < Z < 2.31) = \mathrm{P}(Z < 2.31) - \mathrm{P}(Z < -2.45)$$

$$= 0.9896 - 0.0071 = 0.9825.$$

Approximately 98.25% of the steel rods will lie within the specification limits. [VI.G.2]

95. d; The specification limits are determined externally, usually by management, engineers, or customers. There is no relationship between the specification limits and control limits. The formula in (a) is the natural limits of a process. The formula in (b) is the control limits of a process. The formula in (c) is the definition of the capability index C_p. [VI.G.2]

96. b; USL = 25.85, LSL = 16.85, μ = 20.98, σ = 1.10.

$$C_p = \frac{\mathrm{USL - LSL}}{6\sigma} = \frac{25.85 - 16.85}{6(1.10)} = 1.36$$

$$C_{pk} = \min\left\{\frac{\mathrm{USL} - \mu}{3\sigma}, \frac{\mu - \mathrm{LSL}}{3\sigma}\right\} = \min\left\{\frac{25.85 - 20.98}{3(1.10)}, \frac{20.98 - 16.85}{3(1.10)}\right\}$$

$$= \min\{1.48, 1.25\} = 1.25.$$

Since the actual capability, 1.25, is greater than 1.0, this process is considered capable. [VI.G.3]

97. c; Since the process mean is not centered in the specifications, C_{pk} is used as the measure of capability:

$$C_{pk} = \min\left\{\frac{USL-\mu}{3\sigma}, \frac{\mu-LSL}{3\sigma}\right\} = \min\left\{\frac{19-13}{3(1.5)}, \frac{13-9}{3(1.5)}\right\} = \min\{1.33, 0.89\} = 0.89.$$

Since $C_{pk} < 1$, the process is not capable. [VI.G.3]

98. b; USL = 2.55, LSL = 2.45, $\bar{\bar{X}} = 2.5014$, $\bar{R} = 0.0568$, $n = 7$.

Since the diameter of the steel rod is normally distributed and the process is stable, one can calculate the capability of the process. From the control charts, the process standard deviation is:

$$\hat{\sigma} = \frac{\bar{R}}{d_2} = \frac{0.0568}{2.704} = 0.021.$$

The value for d_2 can be found in Appendix B of *The ASQ Certified Quality Engineer Handbook*, Fifth Edition for $n = 7$. The process is centered in the specifications, so C_p is used as the capability metric

$$C_p = \frac{USL-LSL}{6\sigma} = \frac{2.55-2.45}{6(0.021)} = 0.79.$$

Since the capability is less than 1, the process is not considered capable. [VI.G.3]

99. b; P_p is a performance index that has been recommended for use when the process is not in statistical control. [VI.G.4]

100. c; A treatment combination in experimental design refers to the combination of the levels of each factor assigned to an experimental unit. [VI.H.1]

101. b; The experiment has three factors: temperature, pressure, and time. Each factor is tested at two possible levels. There are four replicates for each treatment combination. [VI.H.1]

102. b; Experimental error is the variation observed in the readings for a particular treatment combination or run. Note that if the number of replications is decreased, the calculation of experimental error is less accurate. [VI.H.1]

103. d; The first step in designing an experiment is to clearly state the objective or goal of the experiment. Defining the goal of the experiment will help with the selection of factors, responses, and measurement method, and selecting the experimental design. [VI.H.2]

104. c; Randomization is the ordering of treatment combinations in a random sequence to reduce the effect of unwanted nuisance or noise factors that may influence the results. When the nuisance factors cannot be controlled during the experiment, randomizing the treatment combinations will help reduce any effects on the response due to the nuisance variables. Confounding occurs when the effect of one factor is indistinguishable from the effect of another factor or factor interactions. Blocking is used when the nuisance variable can be controlled during the experiment, which eliminates its effect on the response. An interaction is the combined effect of two or more independent factors on the response variable. [VI.H.3]

105. b; Because each experimental run takes several hours and the entire experiment will take place over several days, blocking should be used to eliminate the effect of day. Incorporating the day the experimental run was performed will eliminate any effect that day may have had on the response variable. [VI.H.3]

106. c; There are two factors of interest on the yield. Therefore, there are no blocking factors to consider. A factorial design is appropriate for this experiment. [VI.H.4]

107. b; From the ANOVA table, the test hypothesis is whether the main effects and interactions have a significant effect on the response variable. The p-value for the main effects of factors A and B is very small (0.000), as is the interaction between factors A and B. The p-values of the main effect of factor C and the other interaction terms are large; therefore, the conclusion is that they do not have a significant effect on the response. [VI.H.4]

108. d; If the order in which the treatments are carried out is known, a run chart of the responses or residuals can help assess the validity of independence. Any noticeable patterns in the run chart indicate lack of independence. [VI.H.4]

109. a; The total degrees of freedom for a design with a single replicate are $2^k - 1$. For $k = 4$, there are $2^4 - 1 = 15$ degrees of freedom available. There are four main effects (A, B, C, and D), six two-way interactions (AB, AC, AD, BC, BD, and CD), four three-way interactions (ABC, ABD, ACD, and BCD), and one four-way interaction (ABCD). This is a total of fifteen effects, each with one degree of freedom. Therefore, there are no degrees of freedom left over for error. Process variability cannot be estimated in this case. If there were assurance that the higher-order interactions were not significant, the sparsity-of-effects principle could be used and the degrees of freedom pooled for the higher-order interactions into the error degrees of freedom. [VI.H.4]

110. b; The experiment has $2^{8-1} = 128$ runs. Therefore, the design is a half-fractional factorial design. [VI.H.5]

111. a; In a full-factorial experiment, all main effects and all two-factor interactions are fully estimable. In a fractional factorial experiment, while fewer runs are required, effects are confounded with other effects. [VI.H.5]

112. a; Resolution IV designs have main effects confounded with three-factor or higher interactions, and two-factor interactions confounded with two-factor or higher interactions. With a resolution IV design, it is unknown for sure whether a two-factor interaction is truly significant or whether the two-factor interaction is aliased with is significant. Additional experimental runs may be needed to break these aliases. [VI.H.5]

113. c; Using the defining relation I = ABD = ACE = BCDE, all the aliases can be obtained. For factor B,

$$B = B \times I = B \times ABD = AD$$

$$B = B \times I = B \times ACE = ABCE$$

$$B = B \times I = B \times BCDE = CDE$$

Therefore, main effect B is aliased (or confounded) with the two-way interaction AD, the three-way interaction CDE, and the four-way interaction ABCE. [VI.H.6]

114. d; Robustness means resistance to the effect of variation of some factor. The experimenters can control the ingredients (and the amounts) put into the cake mix. They cannot control for differences in oven temperatures in their customers' homes. Therefore, the cake mix should be robust to differences in oven temperature. [VI.H.6]

115. b; The control factors in the experiment are the three ingredients. There are two noise, or uncontrollable, factors in the experiment as well: oven temperature and baking time. There is a significant control-by-noise interaction, indicating that one of the controllable variables interacts with the oven temperature. This indicates that it may be possible to find settings of the controllable factors that are robust to variations in the oven temperature. [VI.H.6]

Part VII

Risk Management

(37 questions)

QUESTIONS

1. Quality risk management is defined as the:
   a. effect of deviations of specifications on reliability objectives.
   b. effect of uncertainty on quality objectives.
   c. coordinated approach for centralized management of risk.
   d. coordination and control of all risky activities of a system.
2. According to ISO 9001:2015, what enables an organization to determine the factors that could cause its processes and its quality management system to deviate from planned results, to put in place preventive controls to minimize negative effects, and to make maximum use of opportunities as they arise?
   a. Risk culture
   b. Risk-based thinking
   c. Risk register
   d. Failure mode and effects analysis (FMEA)
3. Carrying out preventive action to eliminate potential nonconformities, analyzing any nonconformities that do occur, and taking action to prevent recurrence that is appropriate for the effects of the nonconformity are examples of:
   a. fault tree analysis (FTA).
   b. hazard analysis.
   c. risk-based thinking.
   d. vendor risk oversight.

4. A baby is playing near a glass bottle of pesticide. In this situation, what is the bottle of pesticide?

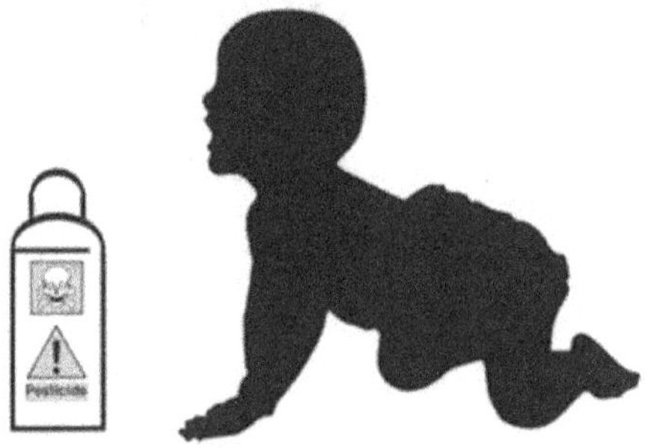

   a. Hazard
   b. Harm
   c. Risk
   d. Impact

5. Risk identification is the process of identifying hazards and sources of risks. When an organization is determining possible risks it:
   a. should include only risks that it has control over.
   b. should include only those risks that are obvious and currently known.
   c. should include all possible risks, using many risk identification tools.
   d. should only come up with a list of no more than three to five risks or hazards.

6. The key stakeholders, strategy, and program objectives, as well as development of roles and responsibilities, are typically identified and agreed on in the:
   a. risk management planning.
   b. risk identification phase.
   c. risk prioritization phase.
   d. risk assessment.

7. In a risk analysis, the likelihood and consequences of risk are combined to determine the risk level for a particular risk. The purpose of risk evaluation is to:
   a. make decisions based on risk level compared to stated risk criteria.
   b. determine options for modifying risks.
   c. decide whether risk levels are acceptable.
   d. evaluate and possibly change the consequences.

8. The rate of aviation fatalities in 1970 was 6.35 per million flights; in 2017, it was 0.39. Which of the following would a company today be failing to understand if it still considered the 1970 rate acceptable?
   a. Internal context
   b. Enterprise context
   c. Cultural context
   d. External context
9. Assessment of organizational objectives and paths to their achievement, prevention of undesired outcomes, and continuous improvement are all goals of:
   a. people management.
   b. technical management.
   c. risk management.
   d. supplier management.
10. As a Quality Manager for new products introduction, the design manager requests that you produce all risk documentation after the design is finalized and frozen to avoid redundant work. Would you follow his recommendation and why?
    a. No, because compliant organizational processes must be followed
    b. Yes, because it doesn't make sense to revise the documents
    c. No, because risk-based thinking begins with the requirements gathering stage
    d. Yes, because risk control measures can best be meaningfully implemented in the later stages
11. Results of risk evaluation must be:
    a. quantitative only.
    b. qualitative only.
    c. subjective.
    d. qualitative or quantitative.

12. Which of the following is a phase of the risk assessment process per ISO 31000:2018?

    a. Risk analysis

    b. Risk planning

    c. Risk monitoring

    d. Risk mitigation

13. Reconsidering objectives and doing nothing are decisions that can be taken after:

    a. risk treatment.

    b. risk monitoring.

    c. risk identification.

    d. risk evaluation.

14. The process of comparing the results of a risk analysis that has been completed to the stated risk criteria put in place is:

    a. risk reduction.

    b. risk identification.

    c. risk evaluation.

    d. risk planning

15. Risk assessment consists of the following:

    a. risk identification, analysis, and control.

    b. risk identification, analysis, and evaluation.

    c. risk identification, reduction, and evaluation.

    d. risk reduction, risk criteria, and evaluation.

16. A risk assessment tool that is based on expert opinion is the:

    a. checklist.

    b. Delphi method.

    c. flowchart.

    d. FMEA.

17. One of the drawbacks or limitations of the FMEA is that:

    a. it cannot be used to identify combinations of failure modes, only single failure modes.

    b. it does not aid in identifying possible root causes.

    c. it is too complicated since it does not require the user to know anything about the process or product.

    d. it cannot be used to prioritize risks.

18. Two components of risk estimation are:

    a. severity and consequence.

    b. probability of occurrence and severity.

    c. probability of occurrence and risk.

    d. severity and risk.

19. Qualitative evaluation involves categories such as catastrophic, critical, major, minor, and negligible, which represent:

    a. probabilities of occurrence.

    b. likelihood of occurrence.

    c. severity.

    d. risk priority number.

20. Consider the following table:

		Probability		
		Low	Medium	High
Severity	Low			
	Medium			
	High			

    This table is an example of a(n):

    a. flowchart.

    b. FMEA.

    c. severity categorization table.

    d. risk matrix.

21. The numerical metric that is an output from an FMEA is the:

    a. risk priority number.

    b. significance level.

    c. capability index.

    d. supplier index.

22. Which of the following risk assessment tools is a top-down approach to failure mode analysis?

    a. FMEA

    b. Failure mode and effects criticality analysis (FMECA)

    c. FTA

    d. Risk management plan (RMP)

23. Which of the following techniques uses guide words and is based on the assumption that risks are caused by deviations from the design plans?

    a. FMEA

    b. FMECA

    c. Hazard analysis and critical control points (HACCP)

    d. Hazard and operability (HAZOP) analysis

24. A new Quality Director is spearheading the establishment of a new Lean Production System at her company. She had previously encountered poor quality products and late deliveries from an overseas supplier and decided to set up a brainstorming session with her team for identifying and recording potential risk sources, causes, and potential consequences of using this overseas supplier to their lean and just in time (JIT) production. What process is being described here?

    a. Risk monitoring

    b. Risk identification

    c. Risk evaluation

    d. Risk communication

25. The purpose of risk control is to:

    a. completely eliminate risk.

    b. completely eliminate all failures.

c. maintain risk at or below an acceptable level.

d. increase probability of occurrence ratings.

26. Risk control requires documentation that addresses monitoring and review of risks and controls put in place. Which of the following would be part of monitoring and review?

    a. Creating the RMP

    b. Obtaining new information to update risk levels and risk criteria

    c. Assigning initial risk ratings

    d. Determining the type of risk management applied

27. A gap may exist when controlling risks if:

    a. there are controls in place to address significant risks.

    b. there is more than one control in place to address a significant risk.

    c. there is no control in place to address a significant risk.

    d. the control in place is sufficient.

28. Residual risk is:

    a. risk that remains after a risk treatment has been implemented.

    b. risk due to a gap between what is believed to be risk and what really is risk.

    c. unknown risk that can never be identified.

    d. risk prior to the implementation of mitigations.

29. Risk reduction and risk acceptance are the two components of:

    a. risk control.

    b. risk audit.

    c. residual risk.

    d. risk reporting.

30. What activity is performed to verify that known risks are being controlled and that risk treatment plans in place are effective?

    a. Risk reduction

    b. Risk audit

c. Risk control

d. Risk reporting

31. Avoiding risk, changing likelihood or consequences, and taking risk to pursue opportunity are actions that are addressed in:

a. ISO 14971:2019.

b. ANSI/ASQ Z1.9.

c. ISO 31000:2018.

d. ISO 9001:2015.

32. Which of the following risk treatment options are available to designers and manufacturers of a hospital bed?

a. Exploit

b. Enhance

c. Mitigate

d. Share

33. A Key Risk Indicator (KRI) is a metric that provides a leading or lagging indication of the current state of risk exposure on key objectives. A manufacturing team assembles circuit boards for transponders. Certain capacitors for the circuit are being found to be faulty upon receipt by incoming quality inspectors, and Return Material Authorization (RMA) forms are often required. The number of RMA forms for a particular supplier is a:

a. leading risk indicator for supplier quality.

b. leading risk indication for supplier timeliness.

c. leading risk indicator for transponder quality.

d. lagging risk indicator for transponder quality.

34. The risk management process step in which the effectiveness of the risk assessment techniques are evaluated is:

a. risk planning.

b. risk treatment.

c. risk monitoring.

d. risk reporting.

35. A medical catheter is approved in the U.S. by the FDA for femoral artery access only. However, physicians are using the device off label for carotid artery access. The clinical department has noticed several publications where carotid access has been performed and successful results published. The company's risk manager states that off-label use is risk that needs to be mitigated. Your marketing team says that this is an opportunity to be capitalized upon. What is the most risk intelligent action to take?

    a. Agree with risk manager that off-label use is a safety risk and should be actively prevented by all means

    b. Acknowledge that off-label use is a risk that is a positive risk and explore an avenue for exploitation using the clinical team's researched data

    c. Agree with marketing team and make plans to approve carotid indication in addition to femoral indication

    d. Don't take any action as current situation is making company money without legal or regulatory liability

36. The main purpose of a risk mitigation plan is to:

    a. eliminate risk.

    b. minimize risk.

    c. accept the risk as is.

    d. maximize risk.

37. If the response to a particular risk is to attempt to minimize its impact, this is known as:

    a. avoidance.

    b. transfer.

    c. mitigation.

    d. denial.

SOLUTIONS

1. b; Quality risk management is the effect of uncertainty on quality objectives. Option c is the definition of enterprise risk management. Option d is part of the definition of operational risk management. [VII.A.1]

2. b; These are elements of risk-based thinking per ISO 9001:2015, clause 0.1. Risk culture is the way of thinking informed by a set of organizational values that determine behavior in face of uncertainty; specifically, the decision making regarding routine and nonroutine risk confrontation and associated risk identification, analysis, evaluation, monitoring, modification and communication. A risk register is a log of assessed and identified risks. FMEA is a form of risk register. [VII.A.1]

3. c; Per ISO 9001: 2015, clause 0.3.3, risk-based thinking embodies preventive action. Fault tree analysis is a tool to identify failures or faults that can lead to a top-level failure event. Hazard analysis is a process to analyze the hazards and their consequences related to product or process under study. Vendor risk oversight involves the vendor or supplier, which is not mentioned here. [VII.A.1]

4. a; A hazard is a potential source of harm or a potential condition that, through a sequence of events, can cause a consequence. A harm is a physical or physiological damage, which here has not been realized yet. Risk here would include scenarios (and their severity and probability) such as the child toppling the bottle or contacting the pesticide. Consequence is the measure of impact – none of which have occurred yet. To classify negative impacts, the term severity is commonly used. [VII.A.1]

5. c; The purpose of risk identification is to identify all possible risks whether current, possible future risks, risks that are not currently under the organization's control, or risks that may occur due to results of an accumulation of factors or steps in the process. [VII.A.1]

6. a; The planning phase is important for setting all ground rules and objectives, and identifying key stakeholders. It may also include selecting key documents and templates for risk analysis. [VII.A.1]

7. a; Risk evaluation is used to make decisions about which risks need to be addressed based on the estimate of metrics identified in risk analysis—combined results of the likelihood of the risk and the consequences. [VII.A.1]

8. d; External context is the environment within which the organization seeks to achieve its objectives. The industry benchmark changed between 1970 and 2017. Cultural context is incorrect and is the environment or situation that is relevant to the beliefs, values, and practices of the culture under study. [VII.A.2]

9. c; Risk management promotes continuous improvement and desired outcomes at the organizational level. The other options may contribute, but risk management provides the overarching umbrella. [VII.B.1]

10. c; It is important to not wait for the final design freeze, but to perform preliminary risk analysis early on. The closer to the requirements-gathering phase the risk analysis is performed, the easier and less expensive it will be to implement risk modification measures. The concept of risk-based thinking is to have not only the designers thinking about potential hazards and their consequences from the first stroke of their pencils but also for the process owners and the finance, marketing, clinical, engineering, operations, and other relevant functions thinking the same way. [VII.B.1]

11. d; Once a risk analysis has been performed, the final evaluation of the results may be quantitative or qualitative, or both. [VII.B.2]

12. a; ISO 31000:2018 divides the risk assessment process into three distinct and sequential phases: risk identification, risk analysis, and risk evaluation. At all levels of risk management, risk assessment is an activating phase that is critical to the risk management process and will require significant resources. [VII.B.2]

13. d; The process of risk evaluation is performed by comparing the estimated risk against the baseline criteria to determine the risk acceptability. It is the exercise of assessing action for estimated risks based on their respective significance to the objectives in terms of concerned product, service, process, or business. ISO 31000:2018 states: "Risk evaluation involves comparing the results of the risk analysis with the established risk criteria to determine where additional action is required. This can lead to a decision to: do nothing further; consider risk treatment options; undertake further analysis to better understand the risk; maintain existing controls; reconsider objectives." [VII. B.2]

14. c; Risk evaluation; once estimates of the evaluation metrics have been calculated, they should be compared to the level of risk or risk criteria already in place to determine whether the level of risk is acceptable. [VII.B.2]

15. b; The three components of risk assessment are risk identification, risk analysis, and risk evaluation. [VII.B.2]

16. b; The Delphi method is based on the idea that group decisions or judgments will be more accurate than those of individuals. [VII.B.2]

17. a; The FMEA cannot identify combinations of failure modes, and failure mode combinations may be significant. [VII.B.2]

18. b; The two components for risk estimation include the probability of occurrence of harm and the severity or potential harm. [VII.B.2]

19. c; The qualitative assessment where the degree of risk or harm is given as catastrophic, critical, and so on, represents the severity or potential harm. [VII.B.2]

20. d; The table represents a risk matrix (also known as a risk assessment matrix). It contains information on the probability of a certain level of risk and the level of severity. [VII.B.2]

21. a; The risk priority number is a numerical result from the FMEA. The RPN is the result of multiplying three values: severity of failure, failure likelihood, and the likelihood of detecting the failure. [VII.B.2]

22. c; FMEA and FMECA are both bottom-up approaches, while FTA is a top-down approach. [VII.B.2]

23. d; HAZOP is the one technique that uses guide words and can be used to identify operability issues and potential hazards that may lead to unacceptable products, processes, or services. [VII.B.2]

24. b; Risk identification is the process of finding, recognizing, describing, and recording risks. Risk identification involves the identification of risk sources, events, their causes, and their potential consequences. Risk reporting and monitoring occurs when real-world data is used to track the risks and or their constituents to ensure that risks are behaving in a manner anticipated during risk treatment and planning. Risk evaluation involves comparing the results of the risk analysis with the established risk criteria to determine where additional action is required. Risk communication involves disseminating the results of various risk management phases. [VII.C.1]

25. c; The main goal of risk control is to maintain the level risk at an acceptably low level. [VII.C.1]

26. b; Obtaining and applying new information should be included in any risk control documentation. [VII.C.1]

27. c; A gap occurs when the control in place is not adequate for the risk, or no control has been put in place for a significant risk. Answer (b) is called a redundancy. [VII.C.1]

28. a; Residual risk is that risk remaining even after a treatment of a known risk has been implemented. There are times when a risk treatment does not eliminate a risk but only minimizes it until the risk is at or below an acceptable level. [VII.C.1]

29. a; Risk reduction focuses on the avoidance of risk, while risk acceptance focuses on the decision to accept risk. Together, they are risk control. [VII.C.1]

30. b; In addition to determining whether known risks are being controlled and risk treatment plans are effective, risk audits can be helpful in identifying new risks. [VII.C.2]

31. d; See Clause 6.1.2 in ISO 9001:2015. [VII.C.3]

32. c; Share, accept, exploit, and enhance options are available only for positive risks. A hospital bed requires safety risk management and will have negative risk treatment options available (i.e., avoid, mitigate, transfer, accept). [VII.C.3]

33. c; The potential defective part is not installed in the transponder. Depending on sampling of supplier lot, the potential exists that a bad part may be used, so the RMA forms would be a leading risk indicator for transponder quality. The supplier has already supplied the defective part, so this would be a lagging risk indicator for supplier quality, which was not provided as an option. [VII.C.4]

34. c; Evaluating the effectiveness of risk management techniques is just one part of monitoring a risk management process. Others include determining whether assumptions made previously when estimating risk are still valid, and whether risk assessment results match or mirror actual results. [VII.C.4]

35. b; It is wise to acknowledge the positive risk and make a well-informed and balanced decision. Preventing would be ignoring positive risk and thus not be risk intelligent. Making plans to approve immediately would be rushing into a decision without risk identification, analysis or evaluation or treatment. Not taking any action would be ignoring the situation and not be risk intelligent. [VII.C.5]

36. b; Risk mitigation includes the act of attempting to minimize the risk's impact on the system, process, or product. [VII.C.5]

37. c; A risk mitigation plan is implemented once risks have been identified. They then must be evaluated to determine whether the risk is worth eliminating or minimizing, or to accept the risk and do nothing about it. [VII.C.5]

Section 2
Additional Practice Problems

Section 2 contains 216 additional practice problems. These questions represent material from each of the seven parts in the *ASQ Certified Quality Engineer* Body of Knowledge. The questions are in a randomized order. Detailed solutions can be found in a section at the end of all the questions.

QUESTIONS

1. In a gage R&R study, the standard deviation associated with repeatability was found to be 0.9214, and the standard deviation associated with reproducibility was found to be 0.5873. What is the standard deviation associated with measurement error?

   a. 0.3314

   b. 1.0927

   c. 1.1939

   d. 1.5087

2. Which of the following is considered attribute data?

   a. Customer wait time

   b. Number of filing errors

   c. Costs of processing applications

   d. Thickness of a component

3. Which of the following is a characteristic of series systems?

   a. The failure of one subsystem results in the failure of the entire system.

   b. They have higher reliability than parallel systems.

   c. More than one subsystem must fail for the system to fail.

   d. They are more complex than parallel systems.

4. A system improvement program based primarily on minimizing variability is called:

   a. Lean Six Sigma.

   b. Six Sigma.

   c. lean.

   d. theory of constraints.

5. A population has a distribution with mean 24 and standard deviation 5.5. What is the approximate standard deviation of the sampling distribution of the sample mean if random samples of size 32 are taken?

   a. $5.5/32$

   b. $5.5/\sqrt{32}$

c. $5.5/\sqrt{24}$

d. 5.5

6. Which of the following involves verifying that the process parameters in place will result in a product or process that meets all requirements under all possible manufacturing conditions?

   a. Installation qualification

   b. Operational qualification

   c. Performance qualification

   d. Software qualification

7. Last month, a local company reported the following quality costs:

Incoming inspection	\$ 60,000
Design support	\$ 12,000
Field trials	\$110,000
Returns by customers	\$ 40,000

   What is the total appraisal cost for last month?

   a. \$60,000

   b. \$65,000

   c. \$72,000

   d. \$170,000

8. A control chart monitoring the viscosity of a product shows that a particular process is out of control. What should the process operator do?

   a. Attempt to fix the problem on their own.

   b. Follow the reaction plan detailed in the control plan of the process.

   c. Brainstorm to determine the cause of the problem.

   d. Nothing; the process will stabilize on its own.

9. An $\bar{X}$ and R chart was prepared for an operation using twenty-five samples with four pieces in each sample. $\bar{\bar{X}}$ was found to be 26.75 and $\bar{R}$ was 8.10. During production, a sample of four was taken and the pieces measured 20, 29, 40, and 25. At the time this sample was taken:

   a. both the average and range were within the control limits.

   b. neither the average nor range were within the control limits.

   c. only the average was outside the control limits.

   d. only the range was outside the control limits.

10. In the third region identified in a bathtub curve, what are the most common type of failures?

    a. Manufacturing process failures

    b. Wear-out failures

    c. Random failures

    d. Burn-in failures

11. Performance measures:

    a. do not need to be achievable.

    b. do not need to be measurable.

    c. should be linked to strategic objectives.

    d. should focus on as many measures as possible.

12. What is the main purpose of an interrelationship digraph?

    a. To organize and summarize ideas generated about a particular problem or issue

    b. To communicate hierarchical relationships between events

    c. To assess a process implementation plan during the initial phases of operations

    d. To visualize and identify causal relationships of a problem or situation

13. A defect was discovered in a laptop battery. How should the company handle the disposition of this defective battery?

    a. Immediately scrap all defective batteries.

    b. Wait for the decision of the Material Review Board (MRB).

c. Begin a product traceability flowchart.

d. Continue production as usual.

14. Suppose that the p-value of a hypothesis test on $\mu_1 - \mu_2$ is 0.23. What conclusions can be made (use significance level of 10%)?

a. Reject H_0; conclude that the two means are not statistically significantly different.

b. Reject H_0; conclude that the two means are statistically significantly different.

c. Do not reject H_0; conclude that the two means are not statistically significantly different.

d. Do not reject H_0; conclude that the two means are statistically significantly different.

15. A component in a car failed three times out of 1,000 trials during testing. If this component fails, the only consequence is that the car fails to cold-start. According to the risk matrix below, what is the risk level of this failure?

		Severity Levels		
		Cold-start failure	Loss of speed during drive	Total loss of power during drive
Occurrence Levels (as % rate of component failure)	5%–10%	Medium	High	High
	1.1%–5%	Low	Medium	High
	0.1%–1%	Low	Medium	High

a. High

b. Medium

c. Low

d. Cannot be determined

16. What is the maximum possible value of an RPN calculated using three 1–10 rating scales?

a. 1

b. 10

c. 100

d. 1000

17. Teams that are formed for the purpose of elevating an existing process or developing a new one are:

    a. self-directed work teams.

    b. work groups.

    c. process improvement teams.

    d. functional teams.

18. A part was inspected using a go/no-go gage. The part mated with both ends of the gage. Did the part pass inspection?

    a. No; the part should be rejected.

    b. Yes; the part should be rejected.

    c. No; the part should be accepted.

    d. Yes; the part should be accepted.

19. What quality tool would be good for conducting a needs analysis when considering a new training program?

    a. Checklists

    b. Histogram

    c. Quality function deployment

    d. Control chart

20. Which phase of the corrective action cycle helps prevent problem backsliding?

    a. Problem identification

    b. Correction

    c. Recurrence control

    d. Effectiveness assessment

21. A team measured the time to failure of twenty components. They plotted the data in a Weibull probability plot, shown below. What conclusions can be drawn from this plot?

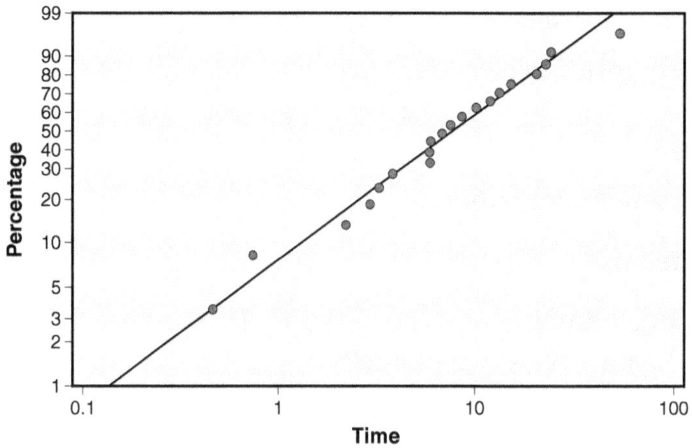

a. The normal distribution is not a reasonable model for the time to failure.

b. The Weibull distribution is a reasonable model for the time to failure.

c. The normal distribution is a reasonable model for the time to failure.

d. Not enough information is provided to draw conclusions.

22. A team investigated the number of patient falls in a hospital. Fifty-six patients were randomly selected. Out of the fifty-six patients, there were six recorded patient falls. Using this information, construct a 90% confidence interval on the true proportion of patient falls in this hospital.

a. (–0.4016, 0.6158)

b. (0, 0.6158)

c. (0.0006, 0.2136)

d. (0.0391, 0.1751)

23. A gage R&R study was performed for a process with ten parts, two operators, and three replicates. The $\bar{X}$ chart below displays information about the gage capability. What conclusions may be drawn from this control chart?

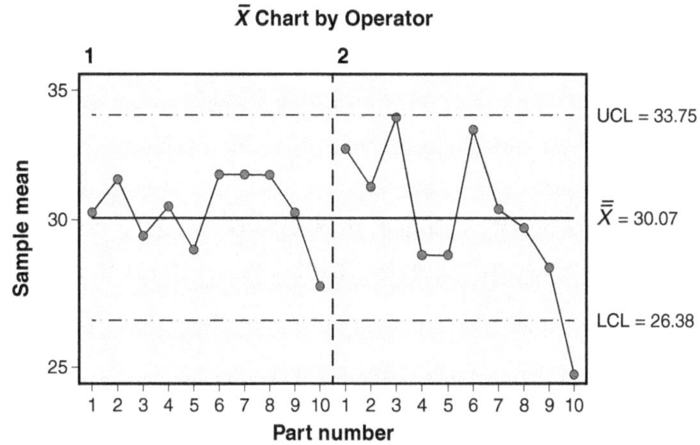

a. The gage is capable of distinguishing between parts.

b. There is a significant difference between operators.

c. The chart reflects variability related to gage reproducibility.

d. The gage is not capable of distinguishing between parts.

24. Consider the product/process development life cycle. The phase in which a working model of the product or process is developed and tested is referred to as the:

a. prototype phase.

b. production phase.

c. concept phase.

d. specification phase.

25. Find the probability that Z, the standard normal distribution, lies between –1.02 and 0.58.

a. 0.1539

b. 0.4349

c. 0.5651

d. 0.7190

26. A lot of size 100 is to be inspected using ANSI/ASQ Z1.4-2003 (R2018). The acceptance quality limit (AQL) is 1.5%. What is the required sample size for a single sampling plan (normal, level II inspection)?

    a. 5

    b. 13

    c. 20

    d. 32

27. The reliability function of a television is defined as R(*t*) = 998/1,000. What is the probability that a randomly selected television from the population fails by time *t*?

    a. 0

    b. 2/1,000

    c. 2/100

    d. 2/100

28. Which quality leader proposed using signal-to-noise ratios to improve robustness of a process?

    a. Deming

    b. Taguchi

    c. Juran

    d. Ishikawa

29. The date of a customer's visit to a store is on which measurement scale?

    a. Nominal

    b. Ordinal

    c. Interval

    d. Ratio

30. Which lean tool results in reductions in process variation, more consistent products, and simplified downstream activities?

    a. Kanban

    b. Single minute exchange of dies (SMED)

c. Overall equipment effectiveness (OEE)

d. Standardized work

31. A control chart was out of control, but the operator failed to adjust the process. What is this called?

a. Common cause variation

b. Special cause variation

c. Overcontrol

d. Undercontrol

32. A new app released for cell phones occasionally slows functionality during customer use. How would the app's developing company classify this defect?

a. Minor defect

b. Major defect

c. Serious defect

d. Critical defect

33. A 3^5 experiment means that there are:

a. three levels of five factors.

b. three independent variables and five replicates.

c. five levels of three factors.

d. five independent variables and three response variables.

34. A company is deciding between two competing sampling plans for a product they receive from a manufacturer. The diagram below displays the operating characteristic (OC) curves for these two sampling plans.

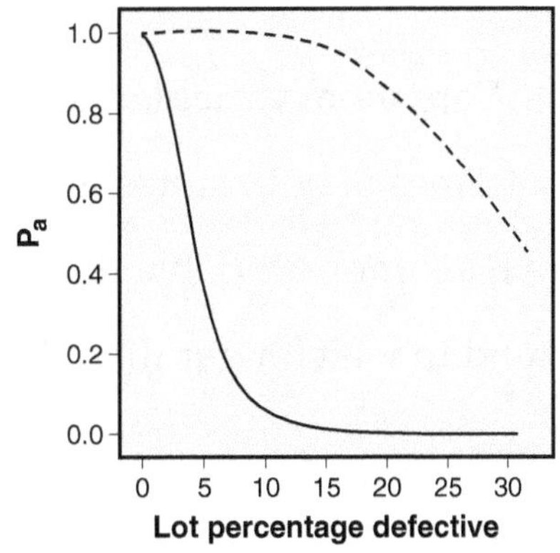

Compared to the solid line, the dashed OC curve:

a. is better for the company.

b. is better for the manufacturer.

c. has a low probability of accepting lots with poor quality.

d. has a low probability of accepting lots with good quality.

35. The failure rate of a machine is 0.0008. Assuming the life distribution of this type of machine is exponential, what is the probability that a randomly selected machine will fail by the mean time to failure (MTTF)?

a. 0.3679

b. 0.6321

c. 0.9992

d. 1,250

36. The side-by-side box plots below show the wait times of customers at two grocery stores. What can be said about these wait times?

a. The wait times at store A are more variable than those at store B.

b. The distribution of wait times at both stores is not bell shaped.

c. The wait times at store B are more variable than those at store A.

d. Customers at store A tend to wait longer than those at store B.

37. A two-way ANOVA was used to test the effects of two factors on a response variable. What conclusions can be drawn from the interaction plot below?

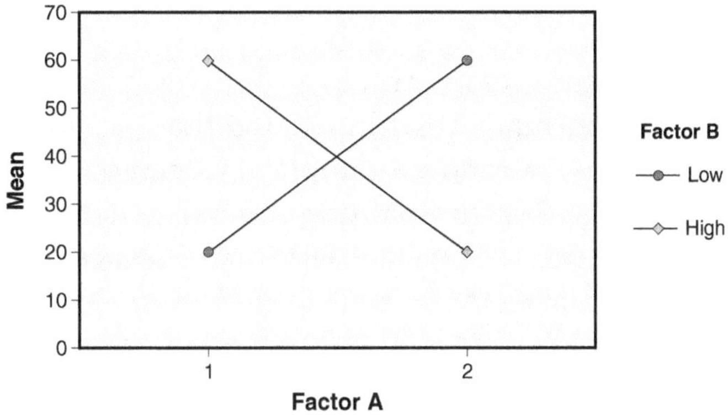

   a. High levels of factor B tend to have smaller response values.

   b. There is not an interaction effect between factors A and B.

   c. There is an interaction effect between factors A and B.

   d. Factors A and B do not have an effect on the response variable.

38. An organization desires to begin rating the capabilities of one of its new suppliers. The process for rating this supplier must include:

   a. rating the supplier's physical location.

   b. rating the supplier's employees' ethnicity.

   c. rating the supplier's financial, manufacturing, and quality systems.

   d. rating the supplier's list of other customers.

39. A quality project team is investigating the root cause of variation in a fish tank manufacturing facility. In what stage of the Define-Measure-Analyze-Improve-Control (DMAIC) process is this team?

   a. D

   b. M

   c. A

   d. I

40. Consider a sampling plan with $n = 80$ and $c = 2$. Given that the percentage nonconforming is 0.025 and a Poisson distribution is appropriate, what is the probability of accepting the lot?

   a. 0.323

b. 0.406

c. 0.677

d. 0.857

41. Tensile strength of cement is of interest to a company. Historical studies indicate that tensile strength is normally distributed and that the standard deviation of tensile strength is 1.25 psi. A random sample of fifteen cement pieces had a mean tensile strength of 124 psi. The company would like to determine if the mean tensile strength is more than 120 psi at the 5% significance level. What is the test statistic for this hypothesis test?

a. –12.39

b. 1.645

c. 1.96

d. 12.39

42. The mean time to failure (MTTF) for a machine is 525 hours. What fraction of machines fail by 1000 hours? Assume the reliability of the machine has an exponential distribution.

a. 0.19%

b. 14.89%

c. 85.11%

d. 99.81%

43. Which of the following components of measurement systems contributes to the precision?

a. Bias

b. Linearity

c. Stability

d. Reproducibility

44. A team suspects that there is a linear relationship between two continuous variables. What is the first thing they should do when examining the relationship between these two variables?

a. Calculate the correlation coefficient.

b. Fit a linear regression model of the two variables.

c. Graph the data in a scatter diagram.

d. Compare histograms of the two variables.

45. If the probability that event A occurs is 0.62, the probability that event B occurs is 0.09, and events A and B are statistically independent, what is the probability that event A and B both occur?

    a. 0.0558

    b. 0.53

    c. 0.71

    d. Not enough information is provided.

46. A system improvement program based on both eliminating waste and reducing variation is called:

    a. Lean Six Sigma.

    b. Six Sigma.

    c. lean.

    d. theory of constraints.

47. In which phase of the corrective action cycle does a team develop feasible solutions to improve a process?

    a. Problem identification

    b. Correction

    c. Recurrence control

    d. Effectiveness assessment

48. Measurement assurance is an important concept in metrology because:

    a. calibration of measuring equipment is critical.

    b. consistent measuring devices provide confidence in the accuracy of the process.

    c. little consideration is often given to a test operator's contribution to measurement error.

    d. random errors in measurement must be taken into account.

49. When there is no control in place to address a significant risk, or the control is insufficient for that particular risk, there is a:

 a. control gap.

 b. risk reduction.

 c. missed risk.

 d. control plan.

50. Increasing the readability of a measuring instrument will lead to increased:

 a. accuracy.

 b. ease of use.

 c. precision.

 d. reproducibility.

51. A safe way to avoid ASQ Code of Ethics violations is to be consistently:

 a. truthful.

 b. opinionated.

 c. prejudiced.

 d. blunt.

52. Which of the following is the activity that may be part of the data transform phase?

 a. Root cause analysis

 b. 5 Why

 c. Splitting tables

 d. Concentration diagram

53. A normally distributed process has specifications 20.35 ± 3.50. $\bar{X}$ and R control charts (based on samples of size four) indicate there is an out-of-control point. The control charts also indicate that $\bar{\bar{X}} = 20.98$ and $\bar{R} = 1.96$. What is the process capability?

 a. 1.01

 b. 1.23

c. 1.45

d. Unknown

54. The scatterplot below displays a relationship between two variables. What is the correlation coefficient of these two variables?

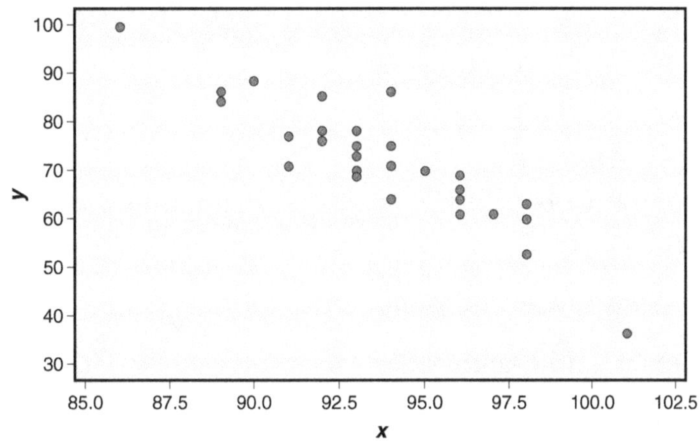

a. –1.82

b. –0.90

c. 0

d. 0.90

55. Evaluation of the capability of equipment, selection of measurement standards, and detailed procedures to carry out measurements are key components of:

a. gage R&R studies.

b. sampling plans.

c. poka-yoke.

d. calibration control systems.

56. Which of the following statements best describes the hazard rate function $h(t)$?

a. It is a measure of the instantaneous rate of failure of a component at time t.

b. It is the probability that a component will survive past time t.

c. It is the fraction of all components in the population that survive past time t.

d. It is the average number of failures through time t.

57. A team would like to estimate the true average tensile strength of cement. They would like to obtain an estimate within 0.05 MPa of the true average tensile strength with 99% confidence. Based on prior information, it is assumed that $\sigma = 0.50$ MPa. What sample size is required to meet these requirements?

    a. 25

    b. 26

    c. 663

    d. 664

58. What are the components of risk as assessed in an failure mode and effects analysis (FMEA)?

    a. Likelihood, occurrence, and detection

    b. Severity, impact, and likelihood

    c. Severity, impact, and detection

    d. Severity, likelihood, and detection

59. The inside dimensions of a part are inspected for conformance to product specifications. Which of the following measuring tools is most appropriate to inspect its dimensions?

    a. Spline

    b. Ring gage

    c. Plug gage

    d. Micrometer

60. A risk manager is focused on reducing the risk of outpatient laparoscopic surgery. After risk mitigations, the number of infections post-surgery has been reduced from 5% to 0.05% over six months but has not been able to be reduced below 0.05% despite further efforts. This level of risk of infection after all possible and imaginable mitigations are applied is known as:

    a. inherent risk.

    b. unacceptable risk.

    c. low risk.

    d. inherent untreatable risk.

61. A factory collected data on the number of nonconforming parts to construct an *np*-chart. Fifteen samples of size 150 were collected. They determined that the average fraction nonconforming was $\overline{p}$ = 0.045. During production, a sample of 150 parts was taken, of which eighteen were nonconforming. At the time this sample was taken:

    a. the sample was within the control limits.

    b. the sample was outside the control limits.

    c. the upper control limit was 0.12.

    d. the lower control limit was 0.

62. What is the effect of a factor that is indistinguishable from the effect of the interaction of two other factors called?

    a. Randomization

    b. Replication

    c. Confounding

    d. Efficiency

63. A company produces a part continuously. Which type of OC curve is appropriate for the proposed sampling plan?

    a. Type A OC curve

    b. Type B OC curve

    c. Type I OC curve

    d. Type II OC curve

64. Consider the following *c*-chart. What conclusions can be drawn from this control chart?

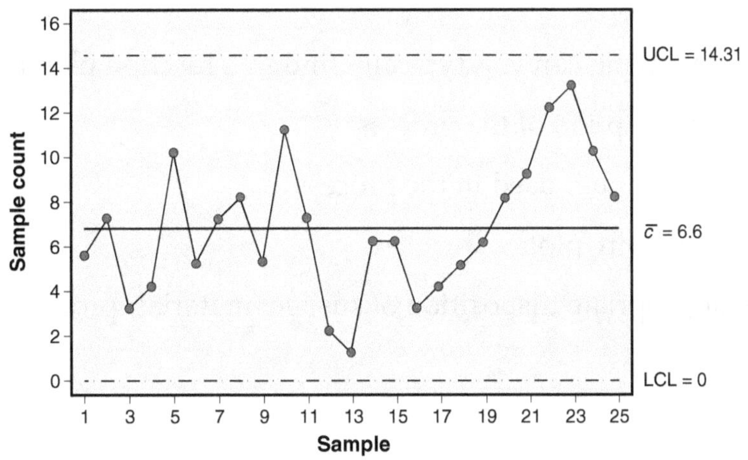

a. The process is out of control.

b. The process is stable.

c. The process is operating within the specifications.

d. The process is operating outside the specifications.

65. In a gage R&R study, the variance associated with repeatability was found to be 0.8842, and the variance associated with reproducibility was found to be 0.2466. What is the total process variance?

a. 0.6376

b. 1.1308

c. 1.4369

d. Not enough information is provided.

66. Which of the following probability distributions is symmetric?

a. Lognormal

b. Exponential

c. Chi-square

d. Normal

67. A company maintained a large stock of supplies in its warehouse. As a result, a sizeable portion of the materials had degraded by the time that they were required. As what type of waste would the company classify this event?

a. Inventory

b. Waiting

c. Overproduction

d. Excess processing

68. Which of the following can you typically find in a reaction plan for a process?

a. A detailed description of the process

b. The machine number used in the process

c. Normal probability plots

d. Steps for appropriate disposition of suspect material/products

69. A team wants to predict a person's cholesterol based on their weight in pounds. The fitted regression equation is Cholesterol = 135 + 0.28 × Weight. What is the predicted cholesterol of a person who weighs 175 pounds?

    a. 0.28

    b. 135

    c. 184

    d. 212.80

70. Common development stages that teams often progress through are known as *forming*, *storming*, *norming*, and *performing*. In the "norming" stage:

    a. individuals begin to shift from personal concerns to the needs of the team.

    b. the team's mission is clarified and specific roles are identified.

    c. team members still think and act as individuals instead of impacting as a team.

    d. the team has matured and is working in the best interest of the team and team goals.

71. A company is releasing a new computer and is developing a flowchart that will include a list of the critical parts in the computer, each part's vendor, and detailed shipping records. The flowchart is an example of:

    a. calibration.

    b. stability.

    c. capability.

    d. traceability.

72. A machine has three identical and independent subsystems, each with a reliability of 0.95. The system design requires that a minimum of two subsystems must function for the system to operate. What is the system reliability of this machine?

    a. 0.8574

    b. 0.9025

    c. 0.9928

    d. 0.9999

73. In an acceptance sampling plan, how does the probability of acceptance change for a given fraction nonconforming as the acceptance number increases? Assume that the sample size and lot size are held constant.

    a. P_a increases.

    b. P_a decreases.

    c. P_a stays constant.

    d. Not enough information is provided.

74. Which of the following is a reason for identifying Critical to Qualities?

    a. To develop a quality process output

    b. To reduce manufacturing costs

    c. To reduce process scrap

    d. To reduce process cycle time

75. A hypothesis test on the variance of the wait times of customers at a store was conducted. You would like to determine if the variance of the wait time is less than 4 minutes^2. A random sample of twenty-two wait times had a variance of 1.75 minutes^2. At the 5% significance level, what is your conclusion from this hypothesis test?

    a. Reject H_0; conclude that the variance equals 4 minutes^2.

    b. Reject H_0; conclude that the variance is less than 4 minutes^2.

    c. Do not reject H_0; conclude that the variance equals 4 minutes^2.

    d. Do not reject H_0; conclude that the variance is less than 4 minutes^2.

76. Accounting for a 1.5σ shift in the mean of a process, how many defective parts per million are produced for a process operating at 6σ quality?

    a. 0.002

    b. 1.5

    c. 3.4

    d. 2,700

77. Suppose an experiment has four factors A, B, C, and D, with four, two, three, and two levels investigated, respectively. How many treatment combinations are in a full-factorial experimental design for these factors?

    a. 4

    b. 11

c. 16

d. 48

78. A manufacturer produces a particular part that is complex. Which of the following measurement tools is most appropriate to inspect its dimensions?

    a. Precision spindle

    b. Coordinate measuring machine

    c. Micrometer calipers

    d. Go/no-go gage

79. Twenty-eight failures occurred after 750 machines were tested for five hours. What is the MTTF for these machines, assuming the reliability of this machine has an exponential distribution?

    a. 0.0075

    b. 0.037

    c. 27

    d. 133

80. Type II error in hypothesis testing corresponds to which of the following in acceptance sampling?

    a. Consumer's risk

    b. Producer's risk

    c. Alpha risk

    d. Unacceptable risk

81. What item below is a benefit of a 5 Why Analysis?

    a. Slower reaction time

    b. Identification of the responsible employee

    c. Future problem prevention

    d. Decreased communication

82. A certain computer company will not build their final product until a customer supplies them with all their desired options. What type of system does this company use?

    a. Pull system

    b. Push system

c. 5S

d. Six Sigma

83. Consider the contingency table presented below. A hypothesis test for independence is conducted to determine whether the number of nonconforming parts produced is independent of time of day.

Time	Conforming parts	Nonconforming parts	Totals
Day	82	18	100
Evening	75	19	94
Overnight	45	14	59
Totals	202	51	253

The value of the test statistic for this hypothesis test is 0.757. At the 10% significance level, what is the conclusion from this hypothesis test?

a. Reject H_0; conclude that type of parts produced and time of day are independent.

b. Reject H_0; conclude that type of parts produced and time of day are not independent.

c. Do not reject H_0; conclude that type of parts produced and time of day are independent.

d. Do not reject H_0; conclude that type of parts produced and time of day are not independent.

84. The failure rate of a machine is 0.0008. Assuming the life distribution of this type of repairable machine is exponential and the MTTR is forty-eight hours, what is the steady-state availability of this machine?

a. 0.037

b. 0.963

c. 1,250

d. 1,298

85. A company's shipment of textbooks to customers became illegible during the printing process due to an ink malfunction. The company must replace each customer's order. How would the company classify this defect?

    a. Minor defect

    b. Major defect

    c. Serious defect

    d. Critical defect

86. A population has a distribution that is highly right-skewed. One hundred samples of size 50 are randomly collected, and the 100 sample means are calculated. What is the approximate distribution of these sample averages?

    a. Normal

    b. Binomial

    c. Uniform

    d. Weibull

87. The number of defects on a TV screen is being examined. Thirty-two TV screens are randomly chosen each day over a 22-day period and examined for defects. The average number of defects is $\bar{c} = 5.68$. What is the upper control limit for a corresponding *c*-chart?

    a. 0

    b. 6.94

    c. 12.83

    d. 22.65

88. What is the recommended first step to take when prioritizing corrective action when many FMEA RPNs are clustered?

    a. Recalculate RPN without occurrence.

    b. Reduce severity.

    c. Improve detection.

    d. Reduce occurrence.

89. To get more accurate measurements, which of the following surfaces is recommended?

    a. Granite

    b. Iron

c. Paper

d. Plastic

90. When driving in the snow and slippery black ice to pick up groceries, which of the following risk treatment options qualifies as risk mitigation?

a. Do not drive at all and so do not pick up groceries until snow melts.

b. Put on snow tires for better traction.

c. Pay your neighbor with a truck to drive and buy groceries for you.

d. Drive in snow and hope for the best.

91. When would inspection be most likely to be considered a value-added activity?

a. For low-cost items

b. For low-risk items

c. To meet legal requirements

d. To evaluate inspector performance

92. Consider the four layers or tiers of a quality manual. Which layer describes the company's commitments, what it stands for, and objectives for quality?

a. First layer/tier: policy statement

b. Second layer/tier: procedures

c. Third layer/tier: work instructions

d. Fourth layer/tier: quality records/results of implementation

93. The voltage of a power supply is of interest to a team. Voltage is assumed to be normally distributed. The voltages of nine randomly selected observations are: 9.66, 8.25, 12.46, 12.35, 11.52, 8.92, 11.00, 11.12, and 11.13. Find the 99% confidence interval on the population standard deviation of the voltage.

a. (0.534, 8.726)

b. (0.783, 12.792)

c. (0.885, 3.577)

d. (9.072, 12.325)

94. Consider the following $\bar{X}$ chart from a normally distributed process. What conclusions can be drawn from this control chart?

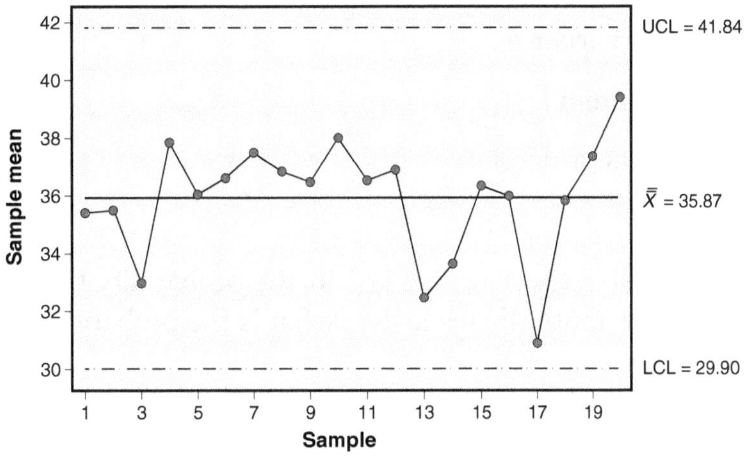

a. The process is out of control.

b. The process is stable.

c. The process is operating within the specifications.

d. The process is operating outside the specifications.

95. The fitted regression equation for two variables x and y is $\hat{y} = -7.5x - 19$. What is the intercept of this line?

a. –19

b. –7.5

c. 7.5

d. 19

96. A manufacturer would like to detect surface defects in a particular part. The surface of this part is made of aluminum. Which of the following testing methods is most appropriate?

a. Liquid penetration testing

b. Magnetic particle testing

c. 100% inspection

d. Coordinate measuring machine

97. Which of the following terms describes the phenomenon in which high-order interactions are negligible and a system is dominated by the main effects and low-order interactions?

    a. Homogeneity of variance

    b. Central limit theorem

    c. Sparsity-of-effects principle

    d. Independence

98. If the probability that event A occurs is 0.40, the probability that event B occurs is 0.13, and A and B are mutually exclusive, what is the probability that either A or B occurs?

    a. 0

    b. 0.052

    c. 0.53

    d. 1.53

99. A hypothesis test is performed at the 5% significance level. The probability of a type II error of the test is 0.15. What is the power for this hypothesis test?

    a. 0.05

    b. 0.15

    c. 0.85

    d. 0.95

100. The acronym NPV stands for:

    a. new product validation.

    b. non-predicted variables.

    c. net present value.

    d. near-perfect variation.

101. The scatterplot below displays the relationship between two random variables. What is the correlation coefficient of these two variables?

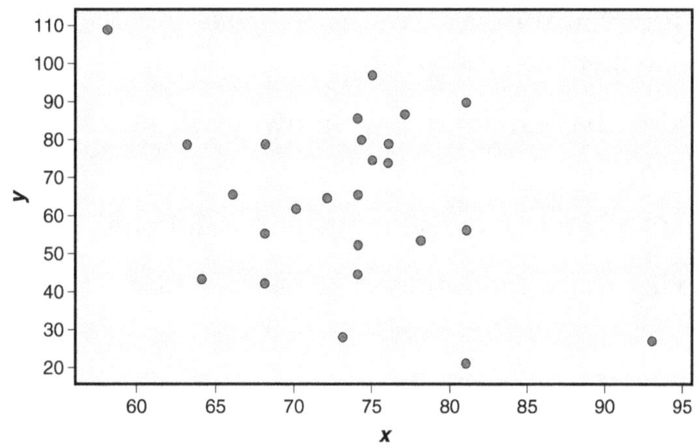

a. –0.91

b. –0.39

c. 0.39

d. 0.91

102. The calibration of a measuring standard for a machine has become faulty. This is an example of:

a. a random error in measurement.

b. a systematic error in measurement.

c. a type I error.

d. a type II error.

103. A specification is given as $10^{+0.002/-0.000}$. The limits for this specification are:

a. 9.998 and 10.000.

b. 9.998 and 10.002.

c. 10.000.

d. 10.000 and 10.002.

104. The waiting times in minutes for eight customers at a bank are: 5, 8, 12, 3, 2, 7, 6, and 5. What is the mode for this sample of waiting times?

a. 3.12 minutes

b. 5 minutes

c. 5.5 minutes

d. 6 minutes

105. The reliability function at time *t* for a lawn mower is R(*t*) = 4,890/5,000. Which of the following statements is true?

a. The probability that a random lawn mower will survive for a time less than *t* is 4,890/5,000.

b. 4,890/5,000 of all lawn mowers in the population fail by time *t*.

c. The probability that a randomly selected lawn mower fails by time *t* is 4,890/5,000.

d. The probability that a randomly selected lawn mower will survive for a time greater than *t* is 4,890/5,000.

106. The dimension of a component has specifications of 10.80 ± 2.50. The process data are normally distributed with mean 10.45 and standard deviation 1.05. Find the actual capability of this process.

a. 0.68

b. 0.79

c. 0.90

d. 1.27

107. According to the ASQ Code of Ethics, confidential information should be:

a. credited.

b. disseminated

c. protected.

d. shared.

108. Philip Crosby is credited with which definition of quality?

a. Fitness for use

b. Conformance to specification

c. Defect-free products

d. Uniformity around a target value

109. A team at a hospital did a project on medication errors. A random sample of medication errors yielded the following results:

Type of error	Number of errors
Incorrect dose	22
Wrong dose	34
Incorrect form	47
Wrong amount	45
Total	**148**

The team used a goodness-of-fit test to determine whether the medication errors occur with equal probabilities. The test statistic was 10.757. At the 5% significance level, what is the conclusion from this hypothesis test?

a. Reject H_0; conclude that the medication errors occur with the same probability.

b. Reject H_0; conclude that the medication errors do not occur with the same probability.

c. Do not reject H_0; conclude that the medication errors occur with the same probability.

d. Do not reject H_0; conclude that the medication errors do not occur with the same probability.

110. In what way does the official definition of the base unit of mass differ from the definitions of other base units?

a. Its definition has been changed several times by the International Bureau of Weights and Measures.

b. Its definition is based on the speed of light.

c. Its definition is in terms of a physical artifact.

d. It is the oldest base unit.

111. If a process is out of control, it is desirable to:

a. not adjust the process.

b. have a small average run length.

c. have a large average run length.

d. stop monitoring the process.

112. Which of the following graphical tools can be used to find the optimal settings of factors in a factorial experiment?

a. Histogram

b. Normal probability plot

c. Contour plot

d. Scatterplot

113. A bottling company's filling process has a normal distribution with mean 24 oz. and standard deviation 0.05 oz. What is the probability that a sample of nine randomly selected bottles will have an average volume less than 23.96 oz?

a. 0

b. 0.0082

c. 0.2119

d. 0.9918

114. Which of the following is outside the normal scope of statistical process control (SPC)?

a. SPC cannot show whether process variation appears to be random.

b. SPC cannot determine whether a process is improving.

c. SPC cannot indicate when a process has nonrandom variation.

d. SPC cannot indicate what caused out-of-control behavior in a process.

115. A certain part has a target length of 5.00 cm. The length of the same type of part was repeatedly measured ten times at three points in time over a two-month period. On day one, the observed average was 5.007; after one month, the observed average was 4.975; after two months, the observed average was 4.824. Which of the following statements is true?

a. There is evidence of nonlinearity in this measurement system.

b. There is evidence of stability issues in this measurement system.

c. The measurement system is not reproducible.

d. The measurement system is not repeatable.

116. Which of the following represents ideal conditions for OEE?

    a. >90% availability, >95% efficiency, >99% quality

    b. >95% availability, >95% efficiency, 95% quality

    c. >99% availability, >95% efficiency, >99% quality

    d. >90% availability, >90% efficiency, >99% quality

117. Which of the following probability distributions is often used to model the first region of the general failure rate model (the bathtub curve)?

    a. Uniform

    b. Exponential

    c. Normal

    d. Weibull

118. Consider the following system where the reliability of each component is displayed within each box. What is the reliability of the overall system?

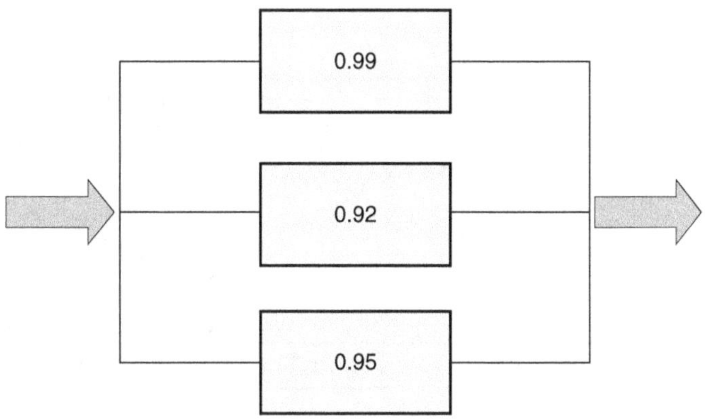

    a. 0.86530

    b. 0.92000

    c. 0.95000

    d. 0.99996

119. An online ordering form for a company will not go through unless the customer has completely filled out all the shipping information and their address is automatically checked as a valid address. This is an example of:

    a. Poka-yoke

    b. Kanban

c. 5S

d. Robust design

120. A quantity derived from sample data that describes the distribution is called a:

a. parameter.

b. statistic.

c. population.

d. critical value.

121. Until the late 1990s, piston engines had a rubber timing belt that synchronized the rotation of crankshaft and the camshaft that ensured that the engine valves opened and closed at the correct time in relation to the pistons. A large risk was the breakage of the rubber belt, which could damage the valves and therefore the engine itself. In the 2000s, the rubber belt was replaced by a metal chain that has minimal wear. The change was plotted on a risk matrix and it was found that the high risk of timing belt was reduced to low risk of timing chain, as shown below. The new risk after risk mitigation of changing the material from rubber to metal is called a:

		Impact Levels				
		Level 1	Level 2	Level 3	Level 4	Level 5
Probability Levels	Level 5	M	M	H	H	H
	Level 4	L	M	H	H	H
	Level 3	L	M	M	H	H
	Level 2	L	L	M	M	M
	Level 1	L	L	L	L	L

Treatment

New Risk associated with metal chain

Original unmitigated risk of rubber belt

a. residual risk.

b. inherent risk.

c. secondary risk.

d. material risk.

122. A disadvantage of using a fractional factorial experiment is that:

   a. it requires more runs.

   b. it is more difficult to analyze.

   c. measurements are less accurate.

   d. effects are confounded.

123. A preventive maintenance policy is recommended when the failure rate is:

   a. decreasing.

   b. constant.

   c. increasing.

   d. unknown.

124. The probability of correctly rejecting a false null hypothesis is called:

   a. power.

   b. type I error.

   c. type II error.

   d. significance level.

125. The scatterplot below displays the relationship between two random variables. What is the correlation coefficient of these two variables?

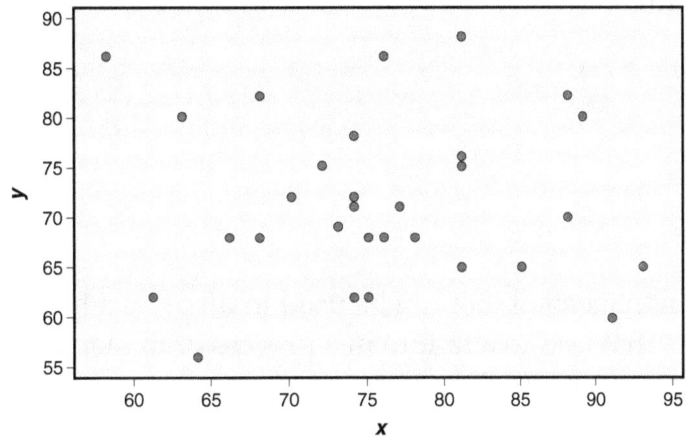

   a. –0.93

   b. 0

   c. 0.93

   d. 1.62

126. A lot size of 50,000 is to be inspected using ANSI/ASQ Z1.4-2003 (R2018). The AQL is 1.0%. What are the required sample size, acceptance number, and rejection number for a single sampling plan (normal, level II inspection)?

a. n = 500, Ac = 10, Re = 11

b. n = 500, Ac = 7, Re = 8

c. n = 800, Ac = 14, Re = 15

d. n = 800, Ac = 10, Re = 11

127. A factory has historically produced defective parts 1% of the time. Samples of fifty independent parts produced in the factory are tested for the presence or absence of a defect. The resulting data (the number of parts that are defective) form which type of distribution?

a. Poisson

b. Hypergeometric

c. Normal

d. Binomial

128. The diameter of a stainless-steel rod is an important quality characteristic. The rod has specifications 1.50 ± 0.08. Data from the process indicate that the distribution is normally distributed, and $\bar{X}$ and R charts indicate that the process is stable. The control charts used a sample of size four and found that $\bar{X} = 1.4983$ and $\bar{R} = 0.0906$. What fraction of steel rods will fall outside the specification limits?

a. 0.0375

b. 0.0689

c. 0.9311

d. 0.9686

129. What quality management tool can be used to discover a hierarchical relationship between events that lead to a failure in a process or to map out the events that lead to a particular outcome?

a. Affinity diagram

b. Activity network diagram

c. Matrix diagram

d. Tree diagram

130. The attribute sampling plan for a specific product utilizes an accept value of 3 and reject value of 7. The inspector discovers six defective items in the current sample. What is the next appropriate action?

 a. Accept the lot, since the number of defective items is less than the Re value.

 b. Reject the lot, since the number of defective items is more than the Ac value.

 c. Draw another sample.

 d. Use a different sampling plan.

131. Active listening by team members is an important skill that is not often easy to attain with new teams. Which of the following is one of the dimensions of active listening?

 a. Listening to music

 b. Confirming

 c. Simultaneous speaking

 d. Being defensive

132. The distribution on which *np*-charts are based is called:

 a. Poisson.

 b. binomial.

 c. normal.

 d. exponential.

133. Consider the contingency table presented below. Given a nonconforming part, what is the probability it was made during the day shift?

Time	Conforming parts	Nonconforming parts	Totals
Day	94	5	99
Evening	87	10	97
Overnight	65	23	88
Totals	246	38	284

 a. 0.02

 b. 0.05

c. 0.13

d. 0.38

134. A random sample of eight pieces of plastic tubing is selected. The inside diameter of the tubing is known to be normally distributed. The inside diameters of the eight pieces of plastic tubing are: 0.183, 0.190, 0.180, 0.189, 0.190, 0.188, 0.186, 0.185.

Find the 95% confidence interval for the population mean of the inside diameter of the plastic tubing.

a. (0.180, 0.190)

b. (0.181, 0.189)

c. (0.183, 0.189)

d. (0.184, 0.188)

135. A quality audit is to be conducted within a particular company. The company has hired an independent contractor to conduct the audit. The contractor is an example of:

a. an internal quality auditor.

b. an external quality auditor.

c. an auditee.

d. a sponsor.

136. The temperature of a chicken entrée prepared in a restaurant was collected in a performance improvement project. The team would like to compare the distribution of the data collected to the target temperature of 165°F. Which of the following quality control tools is most appropriate to use?

a. Histogram

b. Scatter diagram

c. Pareto chart

d. Control chart

137. A double sampling plan is implemented with the following parameters: $n_1 = 75$, $ac_1 = 2$, $re_1 = 5$, $n_2 = 75$, $ac_2 = 4$, $re_2 = 7$. What is the probability that a decision is made (accept or reject the lot) on the first sample if the percent nonconforming $p = 0.015$?

a. 0.2697

b. 0.3211

c. 0.4092

d. 0.5908

138. The MTTF for a model of dishwasher is 115 months. What is the probability of a given dishwasher failing by 115 months of operation? Assume the reliability of the dishwasher has an exponential distribution.

a. 0.009

b. 0.368

c. 0.632

d. 2.718

139. A study on a random sample of sixteen cars was done to evaluate the relationship of age of the car and its cost. The ANOVA table below displays the results of a regression analysis. At the 5% significance level, what conclusions can be drawn from these results?

Source	df	SS	MS	*F*	*p*-value
Regression	1	429.68	429.68	52.53	0.000
Error	14	114.55	8.18		
Total	15	544.23			

a. The age of the car does not have a significant effect on the cost.

b. The intercept of the regression equation, β_0, is equal to 0.

c. The age of the car has a significant effect on the cost.

d. The intercept of the regression equation, β_0, does not equal 0.

140. A team at a factory is monitoring the average number of defects in TV screens. Twenty-eight TV screens are randomly selected each day over a 50-day period. The average number of defects per TV screen was found to be $\bar{u} = 0.214$. What is the lower control limit for a corresponding *u*-chart?

a. 0

b. 0.018

c. 0.410

d. 0.476

141. A benefit of implementing an FMEA is that it:
   a. does not take much time or cost to implement.
   b. must be used with other quality programs such as Six Sigma.
   c. helps identify failures before they occur.
   d. is a reactive tool.

142. In a lean system, what is another name for waste?
   a. Kanban
   b. Kaizen
   c. Seiton
   d. Muda

143. 100% inspection is most appropriate when:
   a. the cost of inspection is high.
   b. destructive sampling is required.
   c. the fraction of nonconforming products is high.
   d. inspection of products causes inspector fatigue.

144. What probability distribution is required for the results of a capability study to be considered valid?
   a. Normal
   b. Binomial
   c. Poisson
   d. Exponential

145. A team decided to change the range of temperature from 150–250°F to 175–225°F in an experimental design. What is a potential disadvantage of changing the range of this factor?
   a. Increasing the experimental error
   b. Missing an important effect
   c. Increasing the cost of the experiment
   d. Increasing the measurement error

146. Consider the histogram below. What shape does this distribution have?

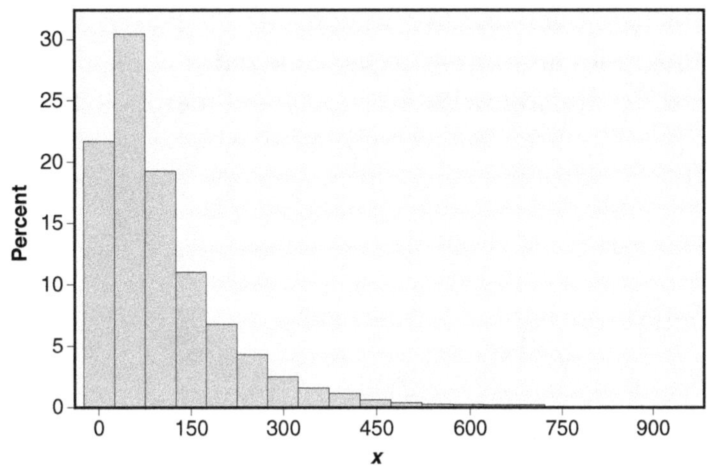

a. Symmetric

b. Right-skewed

c. Left-skewed

d. Bimodal

147. As the degrees of freedom go to infinity, what distribution does Student's t-distribution become?

a. Chi-square

b. Uniform

c. Standard normal

d. F

148. Regulatory requirements, customer needs, and product capabilities are examples of:

a. design control.

b. design review.

c. design outputs.

d. design inputs.

149. Employees at a pharmacy continuously look for ways to make their jobs easier. They color-coded cabinets to better identify materials. They placed check sheets near supplies to keep track of when they needed to be replaced. Finally, they created a document for employees to use to avoid rework when they called customers. Which of the following improvement methods have these employees been using?

    a. Six Sigma

    b. Kaizen

    c. Reengineering

    d. TQM

150. A driver in a Critical to Quality Tree is:

    a. a specification for achieving high quality.

    b. what the customer asks for.

    c. what is needed to achieve what the customer asks for.

    d. a clearly stated need of the customer.

151. Which of the following should be contained in a risk management plan?

    a. Completed FMEA

    b. Risk audit results

    c. Frequency of risk tracking review and reporting

    d. Results of monitoring

152. When compared with attributes sampling plans, variables sampling plans:

    a. provide information on how well a process is operating.

    b. are generally less expensive to implement.

    c. can monitor multiple quality characteristics at once.

    d. are easier to administer in practice.

153. Two methods for filling a bottle are being compared at a factory. The two methods have known standard deviations $\sigma_1 = 0.05$ oz. and $\sigma_2 = 0.04$ oz., respectively. A random sample of $n_1 = 24$ bottles using method 1 and an independent random sample of $n_2 = 25$ bottles using method 2 are collected. The mean volume using method 1 is 15.95 oz., and the mean volume using method 2 is 16.02 oz. Test the hypothesis that the two methods have different mean volumes (use $\alpha = 0.10$). What is your conclusion?

a. Reject H_0; conclude that the two means are not statistically significantly different.

b. Reject H_0; conclude that the two means are statistically significantly different.

c. Do not reject H_0; conclude that the two means are not statistically significantly different.

d. Do not reject H_0; conclude that the two means are statistically significantly different.

154. Which of the following lean tools for helping to reduce changeover times has been attributed to Shigeo Shingo and the Toyota Production System?

a. Cycle time

b. Takt time

c. Standardized work

d. SMED

155. In a test of a component of a machine, 350 components were tested for eight hours and sixty-two components failed. If the mean time to repair these components is three hours, what is the steady-state availability of this component?

a. 0.0221

b. 0.0622

c. 0.9378

d. 45.25

156. Which of the following characteristics of a company's customers is considered continuous data?

a. Age

b. Gender

c. Number of members in the household

d. Hair color

157. Which of the following graphical methods will assess the assumption of normality in the analysis of a designed experiment?

a. Normal probability plot of residuals

b. Scatterplot of residuals versus predicted values

c. Scatterplot of predicted versus actual values

d. Run chart

158. Which of the following statements is true regarding the second region of the bathtub curve?

a. The failure rate is increasing.

b. The failure rate is decreasing.

c. The failure rate is constant.

d. The uniform distribution is a good reliability model.

159. Suppose the number of errors in an invoice at a company has the probability mass function (written in table form) shown below. What is the variance of the number of errors?

Number of errors *X*	0	1	2	3	4
Probability	0.86	0.06	0.04	0.03	0.01

a. 0

b. 0.27

c. 0.5771

d. 0.7597

160. An unknown cause, variable performance from a set standard, and deviation from perception are characteristics of a:

a. Kaizen.

b. solution.

c. control plan.

d. problem.

161. Which of the following actions would result in an ASQ Code of Ethics violation?

a. Obtaining a free conference pass as a presenter

b. Expensing the conference at work as a training event

c. Paying for the conference from personal funds

d. Attending the conference using someone else's badge

162. In a resolution IV fractional factorial experiment, two-factor interactions are confounded with:

   a. main effects.

   b. three-factor or higher interactions.

   c. two-factor or higher interactions.

   d. no other effects.

163. A team is doing a project on errors in customers' files. What type of sampling is appropriate for this project?

   a. Double sampling

   b. Sequential sampling

   c. Multiple sampling

   d. Simple random sampling

164. Which of the following is an assumption required for use and interpretation in regression analysis?

   a. The errors e_i are exponentially distributed with mean zero.

   b. The errors e_i are independent.

   c. The errors e_i are dependent.

   d. The errors e_i have differences in variance.

165. A sample of size 24 is drawn from a population that follows a normal distribution with mean 97 and variance 85. What is the standard error of the sample mean?

   a. 1.88

   b. 3.54

   c. 17.35

   d. 97

166. When assessing process risk through audits, a risk-based audit approach for a design change should include a review of:

   a. evaluation/testing (e.g., human factors/usability, verification, validation, clinical investigations).

   b. supplier evaluation criteria (e.g., self-assessment forms, audits, etc.).

   c. final acceptance criteria.

   d. Nonconforming material reports.

167. A hospital denotes different areas by color (for example, blue, pink, green). In each area, room signs and floors are painted with their designated color. This is an example of:

   a. visual control.

   b. standardized work.

   c. kanban.

   d. waste.

168. The distribution on which *c*-charts are based is called:

   a. Poisson.

   b. binomial.

   c. hypergeometric.

   d. normal.

169. Which of the following tests should be done before performing data analysis?

   a. Scatter test

   b. Matrix test

   c. Normality test

   d. Pareto test

170. A dimension of a component has specifications of 10.80 ± 2.50. The process data are normally distributed with mean 10.45 and standard deviation 1.05. What fraction of components will have this dimension within the specification limits?

   a. 0.0034

   b. 0.0236

   c. 0.9764

   d. 0.9966

171. In one year, a company spent more than $500,000 on warranty issues. This cost would be categorized as:

   a. prevention cost.

   b. appraisal cost.

   c. internal failure cost.

   d. external failure cost.

172. A two-way ANOVA hypothesis test was used to test the effects of two factors on a response variable y. What conclusions can be drawn from the ANOVA table below?

Source	df	SS	MS	F	p-value
A	2	1918.50	959.250	9.25	0.015
B	1	21.33	21.333	0.21	0.666
AB	2	561.17	280.583	2.71	0.145
Error	6	622.00	103.667		
Total	11	3123.00			

a. There is a significant interaction effect between factors A and B.

b. Factor B has a significant effect on the response y.

c. Factor A has a significant effect on the response y.

d. Both factors A and B have a significant effect on the response y.

173. A supplier has received an order from one of their customers. They must complete an order of 750 items in a daytime shift that is ten hours long. The cycle time for this process is forty seconds. Without changing their current production methods, will the supplier be able to meet the customer's demands?

a. No, since cycle time ≤ takt time.

b. Yes, since cycle time ≤ takt time.

c. No, since cycle time > takt time.

d. Yes, since cycle time > takt time.

174. Consider an experiment with seven factors each at two levels. The complete defining relation is I = ABCDF = ABDEG = CEFG. What resolution is this design?

a. Resolution III

b. Resolution IV

c. Resolution V

d. Full-factorial design

175. Paired data collected from a process are: (5.2, 26.7), (6.1, 27.5), (3.2, 24.9), (4.6, 25.5). What is the correlation coefficient for these data?

   a. –0.96

   b. 0.92

   c. 0.96

   d. 21.74

176. Common development stages teams often progress through are known as *forming, storming, norming,* and *performing.* In the *forming* stage:

   a. individuals begin to shift from personal concerns to needs of the team.

   b. the team's mission is clarified and specific roles are identified.

   c. team members still think and act as individuals instead of impacting as a team.

   d. team has matured and is working in the best interest of the team and team goals.

177. What fraction of all rechargeable batteries fail by ten hours if the MTBF is eight hours? Assume the reliability of these batteries has an exponential distribution.

   a. 12.50%

   b. 28.65%

   c. 71.35%

   d. 87.50%

178. The graph below shows the salaries of employees at a company. What measure of central tendency would be most appropriate to describe these data?

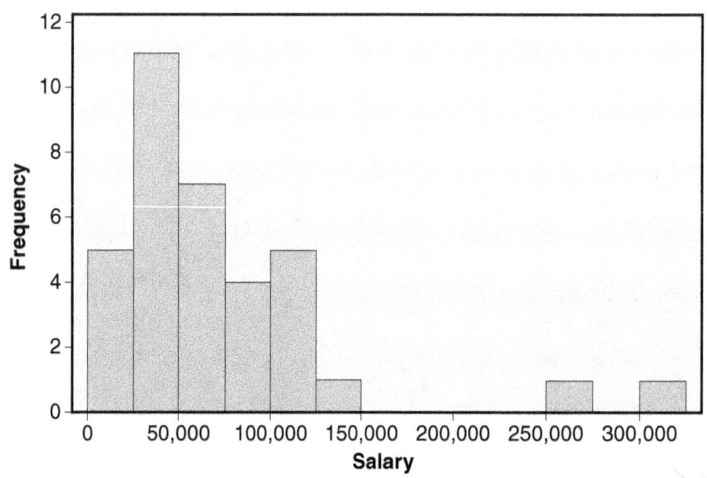

a. Mode

b. Mean

c. Median

d. Standard deviation

179. What is the primary purpose of a process decision program chart?

a. To visualize the relationship between two sets of factors

b. To assess a process implementation plan during the initial phases of operations

c. To manage the schedule of a complex process

d. To evaluate differing options based on a set of criteria

180. A hospital is trying to improve patient flow in its emergency department by tracking patients. Which of the following would help most in this endeavor?

a. Control charts

b. Radio frequency identification

c. OC curves

d. Affinity diagram

181. Which of the following is one of Deming's fourteen points?

a. Institute barriers that stand between the hourly worker and his right to pride of workmanship.

b. Institute modern methods of training on the job.

c. 15% of quality problems are due to the system.

d. 85% of quality problems are due to management.

182. Which of the following statements about rational subgroups is correct?

a. Within-sample variation is maximized.

b. Between-sample variation is minimized.

c. Rational subgrouping is important when constructing control charts.

d. With rational subgroups, there is a smaller probability of variation between successive samples.

183. Gale and Wood (1994) introduced seven tools of customer value analysis. They described different perspectives on what customers believe is of value in a product or service. Which of the following would often be considered a *value* by the customer?

   a. Quality

   b. Inconsistent service

   c. Lack of durability

   d. Difficult to access

184. Consider the following results of a 2^3 full-factorial design with two replicates. Assuming the normality and variance assumptions are valid, which terms have a significant effect on the response variable?

Source	df	SS	MS	F	p-value
A	1	16.0	16.0	3.20	0.1114
B	1	12.0	12.0	2.40	0.1599
C	1	5.0	5.0	1.00	0.3466
AB	1	2.5	2.5	0.50	0.4996
AC	1	1.5	1.5	0.30	0.5988
BC	1	0.5	0.5	0.10	0.7599
ABC	1	0.25	0.25	0.05	0.8287
Error	8	40.0	5.0		
Total	15				

   a. A only

   b. All main effects and two-factor interactions

   c. B only

   d. None of the main effects or interactions

185. A bottling factory claims that the volume of soda filled in bottles is 18 oz. You want to test whether the true mean volume of soda is different than 18 oz. The volume of a random sample of 24 bottles had a mean of 17.98 oz., with a standard deviation of 0.03 oz. Using a significance level of 5%, what is the critical value for this hypothesis test?

a. –3.266

b. ±1.96

c. –±2.069

d. 2.069

186. A tool is designed so that, upon assembly, a component can be attached to the tool in only one position. What type of prevention action does this describe?

a. Fail-safe device

b. Redundancy

c. Magnification of senses

d. Special control device

187. The length of five components was measured by two inspectors (data shown below). Assuming the data follow a normal distribution, find a 95% confidence interval for the difference in the mean length measured by the two inspectors.

Component	1	2	3	4	5
Inspector A	10.2	9.8	10.1	10.3	10.4
Inspector B	10.0	9.9	10.3	10.0	10.3

a. (–0.197, 0.317)

b. (–0.254, 0.374)

c. (–0.367, 0.487)

d. (–0.404, 0.524)

188. The time between arrivals at a store has an exponential distribution with mean four minutes. What is the probability that the time between arrivals is more than five minutes?

a. 0

b. 0.2865

c. 0.7135

d. 3.4903

189. Consider the ANOVA table below for a one-factor experiment with a blocking variable. How many levels are there of factor A and the blocking factor?

Source of variation	df	SS	MS	F
Factor A	5	198.0	49.5	12.05
Block	4	101.75	33.92	
Error	20	79.25	6.60	
Total	29	379.00		

a. 4, 5

b. 5, 4

c. 6, 5

d. 20, 29

190. A group of operators in a semiconductor factory have an idea to improve their workflow. However, upper management has been reluctant to explore this idea. How might this group of operators classify this response?

a. Waiting

b. Excess motion

c. Defect correction

d. Lost creativity

191. Calculate the sample variance of the following observations: 5.3, 8.6, 7.2, 9.1, 7.5, 3.5.

a. 1.928

b. 2.112

c. 3.716

d. 4.459

192. The wait times of thirty consecutive patients at a clinic were recorded. The team determined that $\overline{X} = 17.90$ and $\overline{MR} = 1.75$. What are the upper and lower control limits for the moving range chart?

a. LCL = 0, UCL = 3.29

b. LCL = 0, UCL = 4.65

c. LCL = 0, UCL = 5.72

d. LCL = 13.25, UCL = 22.55

193. A team has collected measurement data from a machine at a factory. To determine whether the data are autocorrelated, which of the following methods could the team use?

a. Trend analysis

b. Goodness-of-fit test

c. Chi-square test

d. Designed experiment

194. Which of the following events is a source of common cause variation?

a. Dull tools

b. Large change in raw materials

c. Slight variation in ambient humidity

d. Broken machine

195. Consider the following $\overline{X}$ chart from a normally distributed process. What conclusions can be drawn from this control chart?

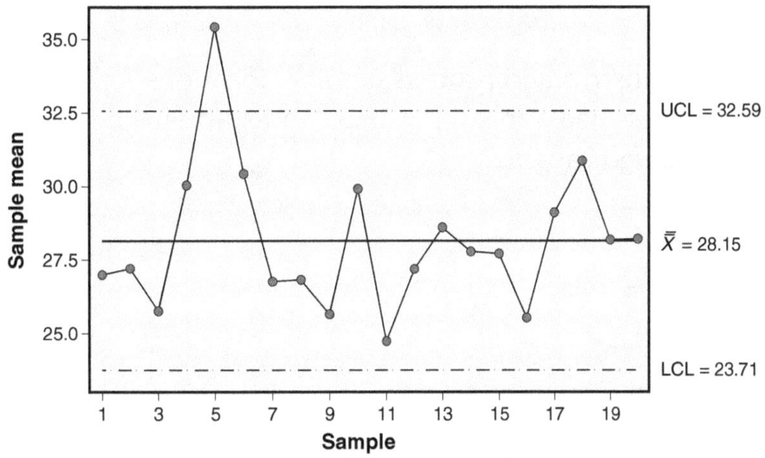

a. The process is out of control.

b. The process is stable.

c. The process is operating within the specifications.

d. The process is operating outside the specifications.

196. Maintaining risk at or below an acceptable level is part of:

a. risk control.

b. risk oversight.

c. risk planning.

d. risk reporting.

197. The test statistic for a hypothesis test for two population variances follows which distribution under the null hypothesis?

a. Normal

b. Binomial

c. χ^2

d. F

198. Which of the following improvement methodologies provides a framework to iteratively improve a process?

a. PDCA

b. TQM

c. Reengineering

d. TOC

199. A stable, normally distributed process has specifications 15.50 ± 1.25. A sample of data from the process had a mean $\mu = 16.50$ and standard deviation $\sigma_1 = 0.10$. Is the process considered capable?

a. Yes, since $C_p = 7.5$

b. Yes, since $C_{pk} = 25$

c. No, since $C_{pk} = 0.83$

d. No, since $C_{pk} = 25$

200. A company has spare parts that need to be evaluated. This would be an example of:

a. prevention costs.

b. appraisal costs.

c. internal failure costs.

d. external failure costs.

201. What is the purpose of a rating scale for consequence or severity?

   a. To quantify risk

   b. To determine the level of impact

   c. To help fill out a risk register/FMEA

   d. To monitor risk

202. Consider the following system where the reliability of each component is displayed within each box. What is the reliability of the overall system?

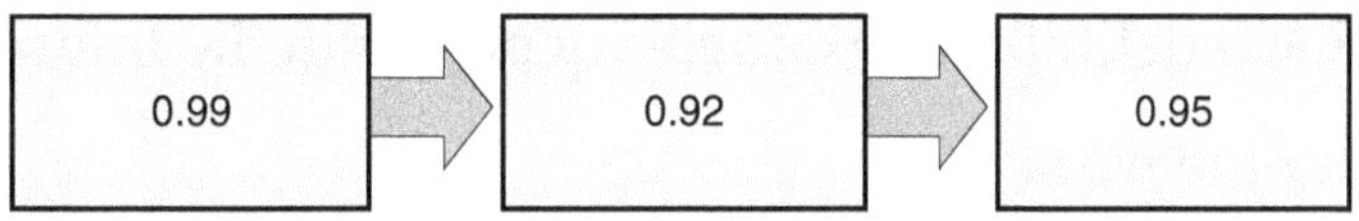

   a. 0.8653

   b. 0.9000

   c. 0.9500

   d. 0.9999

203. Consider the control frame below.

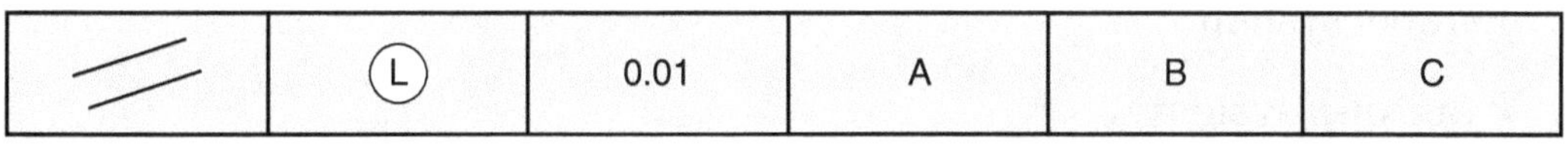

   "B" represents:

   a. primary datum.

   b. secondary datum.

   c. tertiary datum.

   d. modifier.

204. An advantage of a corrective maintenance policy is that it:

   a. minimizes cost.

   b. maximizes run time between repairs.

   c. minimizes risk.

   d. is efficient for catastrophic failures.

205. An online store is reviewing causes for its large number of customer returns. A Pareto chart revealed that the most frequent reason for returns cited by customers was "Item not what I ordered." How would the store classify this type of waste?

    a. Overproduction

    b. Defect correction

    c. Excess movement of materials

    d. Excess processing

206. According to RACI definitions, examples of people in the Informed role are:

    a. people doing the work.

    b. the Accountable person(s).

    c. people who are being asked for advice.

    d. leaders, sponsors, and project champions.

207. A machine has six subsystems and will fail when four or more of these subsystems fail. What type of system design does this machine have?

    a. Series system

    b. Parallel system

    c. *k*-out-of-*n* system

    d. Mixed parallel

208. An experiment has eight factors at two levels each. The experiment has sixty-four runs. What is the experimental design called?

    a. Full-factorial design

    b. Half-fractional factorial design

    c. Taguchi design

    d. One-fourth-fractional factorial design

209. *Criticality* measures

    a. the likelihood of a failure occurring.

    b. the consequences of a failure mode and its frequency of occurrence.

    c. the likelihood of detecting a failure once it has occurred.

    d. the severity of a failure if it occurs.

210. The concept that it is ideal to move as many activities from external to internal as possible to reduce changeover times can be attributed to:

   a. Juran.

   b. Shingo.

   c. Deming.

   d. Ishikawa.

211. The standard deviation of a Poisson distribution is 3.04. What is the mean of this distribution?

   a. 1.74

   b. 3.04

   c. 9.24

   d. 20.91

212. The need for an international standard like ISO 9000 arose:

   a. because national standards are incomplete.

   b. to provide consistency in terms, vocabulary, and requirements.

   c. because industry-specific standards were simply carbon copies of national standards.

   d. because some organizations were more profitable than others.

213. The Malcolm Baldrige National Quality Award emphasizes:

   a. results.

   b. requirements.

   c. procedures.

   d. costs.

214. Last month, a local company reported the following quality costs:

Incoming inspection	$ 60,000
Design support	$ 12,000
Field trials	$110,000

What is the total prevention cost for last month?

a. \$12,000

b. \$77,000

c. \$100,000

d. \$122,000

215. A fault tree analysis (FTA) is a useful tool for prioritizing risks. One of the drawbacks or limitations of the FTA is:

a. it cannot be used to identify combinations of failure modes, only single failure modes.

b. it only uses binary data such as failure/no failure.

c. it is too complicated, since it does not require the user to know anything about the process or product.

d. it cannot be used in certain industries.

216. A risk priority number is an output from a(n):

a. flowchart.

b. QFD.

c. FMEA.

d. expert opinion.

SOLUTIONS

1. b;

$$\sigma^2_{\text{Measurement error}} = \sigma^2_{\text{Reproducibility}} + \sigma^2_{\text{Repeatability}};$$

therefore,

$$\sigma^2_{\text{Measurement error}} = \sqrt{\sigma^2_{\text{Reproducibility}} + \sigma^2_{\text{Repeatability}}} = \sqrt{0.9214^2 + 0.5873^2} = 1.0927.$$

Recall that standard deviation is not additive; variance is. [IV.F]

2. b; *Attribute data* is the term used to describe discrete data in quality control. Wait time, processing costs, and thickness can each be measured on a continuous scale. The number of filing errors can only take on a countable set of numbers, and is considered discrete, or attribute, data. [VI.A.1]

3. a; A series system is made up of n components connected end to end. Therefore, if one component fails, the entire system fails. [III.E.2]

4. b; Lean deals with eliminating waste. Lean-Six Sigma focuses on eliminating waste and minimizing variability. [I.A.1]

5. b; The central limit theorem states that if the sample size n is sufficiently large, then the sample mean $\overline{X}$ follows approximately a normal distribution with mean $\mu_{\overline{X}} = \mu$ and standard deviation $\sigma_{\overline{X}} = \sigma / \sqrt{n}$.

   In this case, $\mu = 24$, $\sigma = 5.5$, and $n = 32$. Therefore,

$$\sigma_{\overline{X}} = 5.5 / \sqrt{32}.$$

   [VI.C.2]

6. b; The goal is to identify limits and action levels that will result in a process or product that meets all of its requirements. [III.D]

7. a; The only appraisal cost is incoming inspection. [II.E]

8. b; Control plans include detailed reaction plans that should be followed when the control method detects a problem in the process. The reaction plan details the steps the operator should take when the process has a problem. [IV.A]

9. d; The summary statistics for the sample are

$$\overline{x} = \frac{20+29+40+25}{4} = 28.50; \; R = 40-20 = 20.$$

   From the information given,

$$\overline{\overline{x}} = 26.75, \; \overline{R} = 8.10, \; n = 4.$$

   The control limits for the $\overline{X}$ chart are:

$$\text{UCL} = \overline{\overline{x}} + A_2\overline{R} = 26.75 + 0.729(8.10) = 32.65$$
$$\text{LCL} = \overline{\overline{x}} - A_2\overline{R} = 26.75 - 0.729(8.10) = 20.85$$

   The control limits for the R chart are:

$$\text{UCL} = D_4\overline{R} = (2.282)(8.10) = 18.48$$
$$\text{LCL} = D_3\overline{R} = (0)(5.50) = 0$$

A_2, D_3, and D_4 can be found in Appendix C of *The ASQ Certified Quality Engineer Handbook*, Fifth Edition for $n = 4$.

Since $\bar{X} = 28.50$ is within the LCL and UCL for the $\bar{X}$ chart, and $R = 20$ is greater than the UCL for the R-chart, only the range was outside the control limits. [VI.F.5]

10. b; The third region of a bathtub curve is the wear-out region, where there is an increasing failure rate over time. Components that rotate or have alternating motions frequently wear out over time. [III.E.3]

11. c; All performance measures should be achievable and aligned with strategic goals. In addition, the focus should be on the "vital few." One should avoid using too many measures or metrics. [I.B.2.c]

12. d; Interrelationship digraphs provide a means to identify sequential or cause-and-effect relationships of a problem or situation. Affinity diagrams are used to organize general thoughts and ideas about a topic. Tree diagrams are used to visualize hierarchical events of a process, and process decision program charts (PDPCs) help to evaluate a process implementation plan at a high level. [V.B]

13. b; When a product defect is discovered, the material review board (MRB) determines what corrective actions to take. [IV.B.4]

14. c; $\alpha = 0.10$ and p-value = 0.23. Using the p-value approach to hypothesis testing, if the p-value $< \alpha$, H_0 is rejected. If p-value $> \alpha$, H_0 is not rejected. Since $0.23 > 0.10$, H_0 is not rejected and the conclusion is that the two means are not statistically significantly different. [VI.D.2]

15. c; For an occurrence rate of 0.3% and cold start failure severity, Low is the correct answer per the provided matrix. [B.2]

16. d; The risk priority number (RPN) is determined by multiplying the three components of risk together: severity, occurrence, and detection. Each of these components is given a score on a 1 to 10 scale. Therefore, the maximum possible value of RPN is $10 \times 10 \times 10 = 1{,}000$. [III.E.4]

17. c; Process improvement teams should also be cross-functional, consisting of people from many different departments/areas within the company (and possibly externally, such as customers and suppliers). [I.D.4]

18. a; A part passes inspection with a go/no-go gage if the part mates with the "go" end and does not mate with the "no-go" end. Since this part mated with both ends, it did not pass the no-go inspection. Therefore, the part should be rejected. [IV.D.1]

19. c; Using a QFD to identify the needs of a training program can provide a more focused approach to design and delivery of the program. [II.F.15]

20. c; In the problem identification phase, the team identifies sources for improvement and develops a clear problem statement. In the correction phase, the team develops feasible solutions to the problem and recommends the best choice to implement. In the recurrence control phase, the team standardizes the solution to ensure that the problem does not reoccur and prevent backsliding. Finally, in the effectiveness assessment phase, the team continues monitoring the process to identify additional opportunities for improvement. [V.E]

21. b; In a Weibull probability plot, the data are plotted against a percentile based on the Weibull distribution. Since the points in this probability plot fall in a straight line, the Weibull distribution is a reasonable model for the time to failure. [VI.A.7]

22. d; In this problem, a "success" is a patient fall. Therefore, $x = 6$ and $n = 56$. The sample proportion is then

$$\hat{p} = \frac{6}{56} = 0.1071.\ \alpha = 0.10,\ Z_{\alpha/2} = Z_{0.05} = 1.645.$$

The corresponding 90% confidence interval is

$$\hat{p} - Z_{\alpha/2}\sqrt{\frac{\hat{p}(1-\hat{p})}{n}} \le p \le \hat{p} + Z_{\alpha/2}\sqrt{\frac{\hat{p}(1-\hat{p})}{n}}$$

$$0.1071 - 1.645\sqrt{\frac{0.1071(1-0.1071)}{56}} \le p \le 0.1071 + 1.645\sqrt{\frac{0.1071(1-0.1071)}{56}}$$

$$0.0391 \le p \le 0.1751.$$

There is 90% confidence that the true proportion of patient falls in the hospital is between 3.91% and 17.51%. [VI.D.1]

23. d; The $\bar{X}$ control chart reflects within-sample variability, which is related to gage repeatability. If the gage is capable, then it should be able to distinguish between parts. Therefore, it would be expected that many of the points would be outside the control limits. In this case, most of the points are within the control limits; this indicates that the gage cannot distinguish between parts. [IV.F]

24. a; In the prototype phase, a working model is created and tested. One goal is to determine whether the product/process will perform as intended. [III.B.1]

25. c; From the properties of the cumulative distribution function, $P(a < Z < b) = P(Z < b) - P(Z < a)$. Therefore, using a standard normal table,

$$P(-1.02 < Z < 0.58) = P(Z < 0.58) - P(Z < -1.02) = 0.7190 - 0.1539 = 0.5651.$$

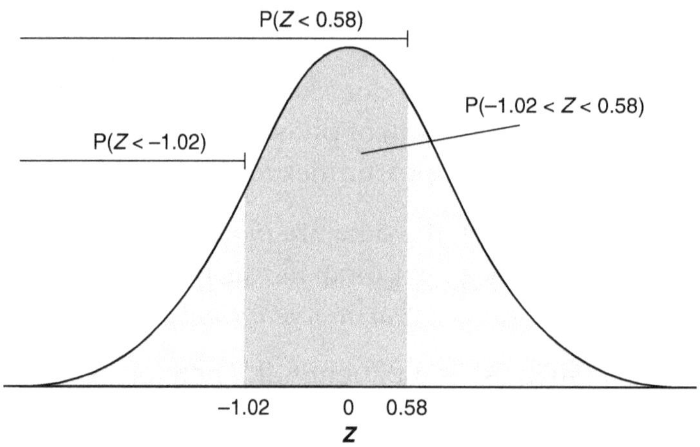

[VI.C.1]

26. d; Given that the lot size is 100, the sample size code letter from the ANSI/ASQ Z1.4-2003 (R2018) sampling tables for normal, level II sampling is F. Using Table II-A and following the down arrow, sampling plan G is referred to and the required sample size is then $n = 32$. [IV.C.2]

27. b; The reliability function at time t is defined as the probability that a randomly selected unit will survive for a time greater than time t. The cumulative distribution function of failure $F(t)$ is the probability that a randomly selected unit fails by time t. Since $F(t) + R(t) = 1$, $F(t) = 1 - (998/1000) = 2/1000$. [III.E.1]

28. b; Genichi Taguchi proposed signal-to-noise ratios to optimize the response variable y and minimize the standard deviation.

If the goal is to maximize the response variable y, the S/N ratio is calculated as

$$S/N = \frac{1}{\bar{y}S}.$$

If the goal of the experiment is to minimize y, the S/N ratio may be calculated as

$$S/N = \frac{1}{\bar{y}S}.$$

If the goal of the experiment is to make y as close to a target value N as possible, the S/N ratio is calculated as

$$S/N=\frac{1}{|\bar{y}-N|S}.$$

[VI.H.5]

29. c; A date is on an interval scale. There are meaningful differences between two dates, but there is no absolute zero. Dates can be added or subtracted. For example, you can calculate the number of days between two dates by subtracting them. However, you cannot meaningfully multiply or divide two dates. [VI.A.2]

30. d; Standardized work means that each activity is done the same way every time. Because each employee performs the activities the same way, having standardized work in practice will help reduce process variation and provide more consistent (and likely better) products. Standardized work can also help simplify activities downstream. [V.D]

31. d; Undercontrol, also called *underadjustment*, occurs when a process operator fails to respond to the presence of special cause variation. *Undercontrol* occurs when a control chart has indicated out-of-control behavior, but the process operator does not adjust the process. [VI.F.2]

32. a; An app that occasionally slows functionality during use would not cause an injury or result in a large economic loss. Therefore, this would be classified as a minor defect. [IV.B.3]

33. a; This notation indicates that there are five independent variables (factors) and each factor is run at three levels. There are a total of $3^5 = 243$ treatment combinations in this experiment. This notation does not give any information on the number of responses or replicates. [VI.H.1]

34. b; The dashed OC curve is better for the manufacturer. The probability of accepting the lot is still very high for large values of percentage nonconforming. The other OC curve would be better suited for the company. [IV.C.1]

35. b; The failure function $F(t)$ is the probability that a randomly selected unit drawn from a population fails by time t:

$$F(t) = 1 - R(t) = 1 - e^{-\lambda t}.$$

Since

$$\lambda = 0.0008,\ \text{MTTF} = \frac{1}{\lambda} = \frac{1}{0.0008} = 1250.$$

Therefore,

$$F\,(\text{MTTF}) = F\,(1250) = 1 - R\,(1250) = 1 - e^{-0.0008(1250)} = 1 - 0.3679 = 0.6321. \text{ [III.E.2]}$$

36. c; The box plot for store A is much tighter than that for store B. This indicates that there is more variability in wait time at store B compared to store A. Note that the distribution of wait times at both stores is approximately symmetric. [VI.A.7]

37. c; An interaction plot can show differences in the mean response for the different levels of factors A and B. This interaction plot indicates that there is a significant interaction between factors A and B. Changing from level 1 to level 2 of factor A, the response is quite different for the low level of factor B. [VI.D.5]

38. c; The supplier's physical location may be of interest to the company but is not necessarily a part of the rating process. It may or may not be included as it is written here. Evaluating or rating on ethnicity would not only be an unethical practice, it is also an illegal one. [I.H.8]

39. c; DMAIC is made up of five phases used in an improvement process: define, measure, analyze, improve, and control. In the analyze phase, the team has already clearly defined the problem and related metrics, and has measured the performance of the process. The team now investigates the root cause of variation in the process or poor performance. [V.C]

40. c; Using the Poisson approximation for the binomial distribution, the probability of acceptance is

$$P_a = P(d \le c) = \Sigma_{d=0}^{c} \frac{(np)^d e^{-np}}{d!} = \Sigma_{d=0}^{2} \frac{(80 \times 0.025)^d e^{-80 \times 0.025}}{d!} = 0.677.$$

Note that $np = 80 \times 0.025 = 2$ is the mean of the binomial distribution, and is therefore the parameter for the Poisson distribution. [IV.C.1]

41. d; This is a hypothesis test for a single population mean μ where the population variance is known. The hypotheses for this test are $H_0 : \mu = 120$ versus $H_a : \mu > 120$. The test statistic is

$$z_0 = \frac{\overline{x} - \mu_0}{\sigma / \sqrt{n}} = \frac{124 - 120}{1.25 / \sqrt{15}} = 12.39.$$

Note that H_0 is rejected if $z_0 > z_\alpha = z_{0.05} = 1.645$. Since $12.39 > 1.645$, H_0 is rejected and the conclusion is that the mean tensile strength is more than 120 psi. [VI.D.2]

42. c; One interpretation of the failure function $F(t)$ is the fraction of all units in the population that fail by time t. Since

$$\text{MTTF} = 525,\ \lambda = \frac{1}{\text{MTTF}} = 1/525.$$

Therefore,

$$F(1000) = 1 - R(1000) = 1 - e^{-\lambda(1000)} = 1 - e^{-\frac{1000}{525}} = 1 - 0.1489 = 0.8511.$$

Approximately 85% of the population will fail by 1,000 hours. [III.E.2]

43. d; The precision of a measurement system consists of *repeatability* and *reproducibility*. Gage bias, linearity, and stability describe the *accuracy* of a measurement system. [IV.F]

44. c; Before calculating the statistics associated with the variables or fitting a regression model, the team should first plot the data in a scatter diagram. The scatter diagram will indicate whether there is a relationship between the two variables. If there is a *linear* relationship, then the team may wish to calculate the correlation coefficient. Comparing the histograms individually will not provide information on how the value of one variable changes with the other variable. [V.A]

45. a; Since A and B are statistically independent,

$$P(A \cap B) = P(A)P(B) = (0.62)(0.09) = 0.0558.$$

[VI.B.3]

46. a; *Lean* deals with eliminating waste. Six Sigma focuses on reducing variation. Lean-Six Sigma focuses on eliminating waste and minimizing variability. [I.A.1]

47. b; In the problem identification phase, the team identifies sources for improvement and develops a clear problem statement. In the correction phase, the team develops feasible solutions to the problem and recommends the best choice to implement. In the recurrence control phase, the team standardizes the solution to ensure that the problem does not reoccur. Finally, in the effectiveness assessment phase, the team continues monitoring the process to identify additional opportunities for improvement. [V.E]

48. c; Measurement assurance includes information on the contribution of error by technicians or a laboratory's procedures. Calibration is typically the focus when testing an instrument, and little consideration is given to errors made by operators or the environment in which it's used. [IV.E]

49. a; When the control in place is not adequate for the risk, or no control has been put in place for a significant risk, it is called a *gap* or *control gap*. [VII.C.1]

50. b; *Readability* is defined as the ease of reading the instrument scale while measuring some component. The readability of the equipment, however, will not necessarily result in more accurate results in measurement. [IV.E]

51. a; Section A of the ASQ Code of Ethics is, "Act with Integrity and Honesty" which includes the statement, "Be truthful and transparent in all professional interactions and activities." [I.C]

52. c; The Data Transform phase's main goal is preparing data to fit the parameters of another system. Transformations include: data sorting and filtering to get rid of irrelevant items, de-duplicating and cleansing, translating and converting, removing or encrypting to protect sensitive information, or splitting or joining tables.

53. d; The control chart indicated there was an out-of-control point. This means that the process is not stable; there is special cause variation present in the system. Therefore, the capability of the process is unknown. [VI.G.2, VI.G.3]

54. b; The sample correlation coefficient measures the strength of the linear relationship between two variables. The scatterplot shows that as x increases, y decreases. This indicates that there is a negative correlation. The correlation coefficient only takes values between –1 and +1. The closer the value is to –1 or +1, the stronger the linear relationship, so there is a strong negative correlation between these two variables. [VI.E.2]

55. d; Typical calibration control systems include the following procedures: evaluation of equipment to determine its capability, identification of calibration requirements, selection of standards to perform calibration, selection of the methods to carry out the measurements for the calibration, and more. [IV.E]

56. a; The hazard function is defined as

$$h(t)=\frac{f(t)}{R(t)}=\frac{\text{\# of failures per unit time}}{\text{\# components tested per unit time}}.$$

Therefore, one interpretation of the hazard function is an instantaneous rate of failure at a given time t. [III.E.1]

57. d; As stated in the problem, the estimate should be within 0.05 MPa of the true average. This means the estimate is allowed to be at most 0.05 MPa less than the average, or at most 0.05 MPa greater than the true average. The margin of error is E = 0.05. Furthermore,

$$\sigma = 0.50,\ \alpha = 0.01,\ Z_{\alpha/2} = Z_{0.005} = 2.576.$$

The minimum sample size needed is

$$n \geq \left(\frac{\sigma Z_{\alpha/2}}{E}\right)^2 = \left(\frac{0.50(2.576)}{0.05}\right)^2 = 663.58.$$

Therefore, the minimum sample size of cement would be $n = 664$. [VI.D.1]

58. d; The risk associated with each potential failure mode is assigned a risk priority number (RPN). The RPN is made up of three components: severity, occurrence, and detection. [III.E.4]

59. c; Plug gages are typically used to assess conformance of inside dimensions of a part. Ring and snap gages are used to assess outside dimensions. Splines are typically used to check special features. Micrometers are used for linear measurements. [IV.D.1]

60. d; The risk remaining after all mitigations is the one that cannot be treated. Note also that residual risk is the risk after mitigations. In this unique case, since all possible mitigations have been applied, residual risk is equal to inherent untreatable risk. Residual risk is a function of level of treatment of mitigative effort. Inherent risk is risk before mitigations. [VII.C.5]

61. b; From the information provided, $\bar{p} = 0.045$, $m = 15$, $n = 150$. The new sample number nonconforming is $x = 18$.

    The control limits for the *np* chart are

$$\text{UCL} = n\bar{p} + 3\sqrt{n\bar{p}(1-\bar{p})} = 150(0.045) + 3\sqrt{150(0.045)(1-0.045)} = 14.36$$

$$\text{LCL} = n\bar{p} - 3\sqrt{n\bar{p}(1-\bar{p})} = 150(0.045) - 3\sqrt{150(0.045)(1-0.045)} = 4.211$$

    Since $x = 18$ lies above the UCL of the *np* chart, the sample is outside the control limits. [VI.F.5]

62. c; *Confounding* occurs when the effect of one factor is indistinguishable from the effect of another factor or factor interactions. An *interaction* is the change in response when two or more factors are interdependent. *Randomization* is the ordering of treatment combinations in a random sequence to reduce the effect of nuisance variables on the response. *Efficiency* in experimental design refers to an experiment that generates the maximum information while using a small number of runs and resources. [VI.H.3]

63. b; Type B OC curves are used to evaluate a sampling plan of a continuous process. These curves are constructed using either a binomial or Poisson distribution, depending on the scenario. Type A OC curves are used on a lot-by-lot basis for a product of a continuous process. [IV.C.1]

64. a; There is a sequence of eight points, all increasing beginning with the sixteenth data point. One of the sensitizing rules to detect out-of-control behavior is six points in a row, all increasing or all decreasing. Note that a control chart does not provide information on whether the process is operating within the specification limits. A process could be out of statistical control, but still operating within the specification limits. Similarly, a process could be in control, but not be within the specification limits. [VI.F.6]

65. d; The total variability is composed of the variability associated with the gage or measurement tool in addition to the part-to-part variability. Information on the part-to-part variability is not provided here, so there is not enough information to determine the total variability. Recall that

$$\sigma^2_{\text{Total}} = \sigma^2_{\text{Gage}} + \sigma^2_p = \sigma^2_{\text{Reproducibility}} + \sigma^2_{\text{Repeatability}} + \sigma^2_p,$$

where σ^2_p denotes the part-to-part variability. [IV.F]

66. d; The lognormal, exponential, and chi-square distributions are all skewed to the right. They also can assume only positive values. [VI.C.1]

67. a; This is an example of waste due to inventory. The company has more inventory in storage than is necessary. As a result, it must pay to store these materials, which have no value to the customer. In addition, the company's high inventory also leads to additional losses when some of the products degrade before use. [V.D]

68. d; A reaction plan typically contains four elements: containment, diagnosis, verification, and disposition. The reaction plan should therefore include specific directions for the operators on how to dispose of material that was contained when a problem was detected in the process. Disposition may include scrap, rework, sort, use as-is, or return to vendor. [IV.A]

69. c; By plugging in the given value of Weight = 175 into the regression equation, the cholesterol can be predicted. Cholesterol = 135 + 0.28 × (175) = 184. [VI.E.1]

70. a; In the *norming* phase, interpersonal conflicts have less impact on team dynamics and work. [I.D]

71. d; A material flow or traceability flowchart should include the critical components and their part numbers, a list of the vendors for each component, information on any internally manufactured part (including the date of manufacture), and detailed, computerized shipping records. [IV.B.1]

72. c; The machine has a k-out-of-n system design where $k = 2$ and $n = 3$. The system reliability of a k-out-of-n system is

$$R_s(t) = \Sigma_{i=k}^{n} \binom{n}{i} p^i (1-p)^{n-i},$$

where $p = 0.95$. Therefore, the system reliability is

$$\begin{aligned} R_s(t) &= \binom{3}{2}(0.95)^2(0.05)^1 + \binom{3}{3}(0.95)^3(0.05)^0 \\ &= 3(0.95)^2(0.5)^1 + 1(0.95)^2(0.05)^1 \\ &= 0.1354 + 0.8574 \\ &= 0.9928. \end{aligned}$$

[III.E.2]

73. a; Recall that

$$\mathrm{P_a} = \Sigma_{d=0}^{c} \binom{n}{d} p^d (1-p)^{n-d}.$$

Therefore, as c increases, the probability of acceptance will increase for a given fraction nonconforming and sample size. Consider, for example, an infinite lot size, a sample size $n = 15$, and fraction nonconforming $p = 0.10$. For acceptance numbers c, the following probability of acceptance is

Acceptance number (c)	P_a
0	0.2059
1	0.5490
2	0.8159
3	0.9444

[IV.C.1]

74. a; Critical to Qualities are useful for the development of a quality process output. [III.B.1]

75. b; The null and alternative hypotheses for this hypothesis test are

$$\mathrm{H_0} : \sigma^2 = 4 \text{ versus } \mathrm{H_a} : \sigma^2 < 4.$$

The level of significance is $\alpha = 0.05$. This is a left-tailed test, so the null hypothesis is rejected if the test statistic is less than

$$\chi^2_{1-\alpha,k} = \chi^2_{0.95,21} = 11.591.$$

The test statistic is

$$\chi^2_0 = \frac{(n-1)s^2}{\sigma^2_0} = \frac{(22-1)1.75}{4} = 9.1875.$$

Since $9.1875 < 11.591$, the null hypothesis is rejected, and it is concluded that the alternative hypothesis is true. At the 5% significance level, there is sufficient evidence to conclude that the variance of wait time is less than 4 minutes^2. [VI.D.2]

76. c; For a process operating at Six Sigma level quality, there are 3.4 defective parts per million. Not accounting for the 1.5α shift from the mean, the process would have 0.002 defective parts per million. [V.C]

77. d; A full-factorial design is an experiment such that all treatment combinations of the factors are carried out. Therefore, the number of treatment combinations is $4 \times 2 \times 3 \times 2 = 48$. [VI.H.1]

78. b; Coordinate measuring machines are used for dimensional quality control for complex parts. They allow measurements to be completed within minutes. Precision spindles are used to make roundness measurements, micrometer calipers are used to take linear measurements, and go/no-go gages are used to assess specification conformance. [IV.D.1]

79. d; The failure rate, λ, is the number of failures divided by the total test time. Therefore,

$$\lambda = 28 \ / \ (750 \times 5) = 0.0075.$$

The MTTF is $1/\lambda \approx 133$, assuming the life distribution is exponential. [III.E.2]

80. a; A type II error in hypothesis testing is the probability of failing to reject the null hypothesis when in fact the null hypothesis is false. This corresponds to the consumer's risk in acceptance sampling: the probability of accepting a lot that does not meet the lot tolerance percent defective (LTPD) level. [IV.C.1]

81. c; The purpose of a 5 Why root cause analysis is to identify the system or process that caused the issue. Fast reaction time, future problem prevention, and improved communication are benefits of a 5 Why root cause analysis. [V.E.]

82. a; A pull system is emphasized in a lean enterprise. In a pull system, an action will initiate the next step in the process occurring. In this case, the customer's order initiates the completion of the computer. Using a pull system eases the burden of making accurate forecasts of customer demand and helps to reduce unused inventory. [V.D]

83. c; The test statistic for the test of independence follows a chi-square distribution with $(r-1)(c-1)$ degrees of freedom where r represents the number of levels of the first factor and c represents the number of levels of the second factor. H_0 is rejected if

$$\chi_0^2 > \chi^2_{\alpha,(r-1)(c-1)} = \chi^2_{0.10,2} = 4.605.$$

Since

$$\chi_0^2 = 3.359 < 4.605,$$

H_0 is not rejected, and it is concluded that the type of part produced (conforming or nonconforming) and time of day are independent. [VI.D.6]

84. b; The failure rate for this machine is $\lambda = 0.0008$, which means that the MTBF is

$$\frac{1}{\lambda} = \frac{1}{0.0008} = 1250 \text{ hours.}$$

Since MTTR = 48 hours, the steady-state availability of the machine is

$$A = \frac{\text{MTBF}}{\text{MTBF}+\text{MTTR}} = \frac{1250}{1250+48} = 0.9630.$$

[III.E.2]

85. c; While the illegible textbooks will not cause injury, they do represent a large economic loss, as the company must replace all of the defective books. The company would classify this as a serious defect. [IV.B.3]

86. a; The central limit theorem states that the distribution of sample averages tends toward a normal distribution for large sample sizes. Samples size of greater than thirty or forty is the typical rule of thumb for large n. [VI.A.6, VI.C.2]

87. c; From the information provided,

$$\bar{c} = 5.68,\ n = 32,\ m = 28.$$

$$\text{UCL} = \bar{c} + 3\sqrt{\bar{c}} = 5.68 + 3\sqrt{5.68} = 12.83.$$

[VI.F.5]

88. a; Note that ranking RPNs works effectively to prioritize corrective action when there is "comfortable" separation between the RPN values. When these values are clustered, the recommended strategy is to eliminate the occurrence to reorder the RPNs, then reduce the severity, reduce the occurrence, and improve detection. [III.E.4]

89. a; A granite surface plate provides better hardness compared to iron surface plates and other surfaces. In addition, granite surface plates are more resistant to corrosion and are less responsive to changes in temperature than iron surface plates. [IV.D.1]

90. b; Putting on snow tires is a mitigation. Not driving at all is risk avoidance. Using the neighbor is risk transfer. Hoping for the best is risk acceptance. [C.3]

91. c; There are situations where inspection may be considered non-value-added and other situations where it is considered value-added. Inspection of low-cost items, for example, tends to be considered a non-value-added activity. However, inspection of a higher-risk product, or one that is required to be tested by law, could be considered value-added. [V.D]

92. a; The policy statement is the first layer in the manual because it not only provides information about the company's commitment, but policy statements are often provided for each of the applicable standards. [II.B.1]

93. c;

$$\overline{x} = \frac{\Sigma_{i=1}^{n} x_i}{n} = \frac{9.66+8.25+12.46+12.35+11.52+8.92+11.00+11.12+11.13}{9}$$

$$= 10.712 \text{ V}$$

$$s^2 = \frac{\Sigma_{i=1}^{n}(x_i - \overline{x})^2}{n-1} = \frac{(9.66-10.712)^2 + \ldots + (11.13-10.712)^2}{9-1} = \frac{4.687}{8} = 2.149 \text{ V}^2$$

The 100(1 – α)% two-sided confidence interval on the population standard deviation σ is given by

$$\sqrt{\frac{(n-1)s^2}{\chi^2_{\alpha/2,k}}} < \sigma < \sqrt{\frac{(n-1)s^2}{\chi^2_{1-\alpha/2,k}}}.$$

$$n = 9,\ s^2 = 2.149,\ \alpha = 0.01,\ k = n-1 = 8,\ \chi^2_{0.005,8} = 21.955,\ \chi^2_{0.995,8} = 1.344.$$

$$\sqrt{\frac{(9-1)2.149}{21.955}} < \sigma < \sqrt{\frac{(9-1)2.149}{1.344}}$$

$$0.885 < \sigma < 3.577.$$

There is 99% confidence that the true population standard deviation of the voltage of the power supply lies between 0.885 V and 3.577 V. [VI.D.1]

94. a; There are nine points in a row above the centerline beginning with the fourth data point. One of the sensitizing rules to detect out-of-control behavior is nine points in a row on the same side of the centerline. Note that a control chart does not provide information on whether the process is operating within the specification limits. A process could be out of statistical control, but still operating within the specification limits. Similarly, a process could be in control, but not be within the specification limits. [VI.F.6]

95. a; The intercept is the value of y when $x = 0$. Therefore, the intercept is –19. [VI.E.1]

96. a; Because the surface of the part is made of aluminum (a nonmagnetic material), liquid penetration testing is most appropriate to detect surface defects. Note that magnetic particle testing cannot be used because aluminum is a nonmagnetic material. [IV.D.2]

97. c; The sparsity-of-effects principle is an assumption that is often made that higher-order interactions (orders higher than two-factor interactions) do not have a large effect on the response. This principle is useful when full-factorial designs are conducted with only one replicate. [VI.H.4]

98. c; Since A and B are mutually exclusive, events A and B cannot occur simultaneously. Therefore, $P(A \cap B) = 0$. Then $P(A \cup B) = P(A) + P(B) = 0.40 + 0.13 = 0.53$. Note also that probabilities only take on values between 0 and 1. [VI.B.3]

99. c; Power = 1 – P(Type II error) = $1 - \beta = 1 - 0.15 = 0.85$, or 85%. Note that the significance level is α = P(Type I error) = 0.05. [VI.D.2]

100. c; NPV is short for net present value. [I.B.2.c]

101. b; The scatterplot shows that as x increases, y tends to decrease, indicating a negative correlation. The points are not closely clustered together, indicating a weak negative correlation. Therefore, –0.39 is the appropriate answer. [VI.E.2]

102. b; Errors in measurement are classified as *random* errors (accidental fluctuations that cannot be predicted) or *systematic* errors (for example, operator error, calibration error). Faulty calibration is classified as systematic error. [IV.E]

103. d; This is a unilateral tolerance with variation in only one direction. [III.C]

104. b; The mode is the observation that occurs most often in the sample. In the sample data, a five-minute wait time occurs more than any other observation. [VI.A.6]

105. d; The reliability function is defined as the fraction of all components that have survived for a time greater than *t*. Therefore, the probability that a randomly selected lawn mower will survive longer than *t* is 4890/5000. Note that the cdf of the failure time [$F(t) = 1 - R(t)$] can be interpreted as the probability that a randomly selected unit fails by time *t* or is the fraction of all units in the population that fail by time *t*. [III.E.1]

106. a; The actual capability is the capability measure C_{pk}.

$$C_{pk} = \min\left\{\frac{USL-\mu}{3\sigma}, \frac{\mu-LSL}{3\sigma}\right\} = \min\left\{\frac{13.3-10.45}{3(1.05)}, \frac{10.45-8.3}{3(1.05)}\right\}$$

$$= \min\{0.90, 0.68\} = 0.68.$$

Since the capability is less than 1, the process is not considered capable. [VI.G.3]

107. c; The ASQ Code of Ethics says, "Ensure the protection and integrity of confidential information." [1.C]

108. b; Crosby is credited with conformance to specification. Juran is credited with "fitness for use" and Genichi Taguchi is credited with "uniformity around a target value." "Defect-free product" was never a definition of quality. [I.A.1]

109. b; The team would like to test whether the medication errors occur with equal probabilities. Therefore, the null and alternative hypotheses are

$$H_0 : p_1 = p_2 = p_3 = p_4 = 0.25 \text{ versus } H_a : p_i \neq 0.25 \text{ for at least one } i = 1, 2, 3, 4.$$

The significance level is $\alpha = 0.05$. H_0 is rejected if

$$\chi_0^2 \geq \chi_{\alpha,k-1}^2 = \chi_{0.05,4-1}^2 = 7.815.$$

$$\chi_0^2 = \sum_{i=1}^{k} \frac{(o_i - E_i)^2}{E_i}, \quad \text{where } E_i = (148)(0.25) = 37 \text{ for } i = 1,2,3,4.$$

$$\chi_0^2 = \frac{(22-37)^2}{37} + \frac{(34-37)^2}{37} \frac{(47-37)^2}{37} + \frac{(45-37)^2}{37} = 10.757.$$

Since 10.575 > 7.815, H_0 is rejected, and the conclusion is that the medication errors do not occur with equal probabilities. [VI.D.4]

110. c; The kilogram, the base unit of mass, is the only base unit defined in terms of a physical artifact, which is made of a platinum-iridium alloy and kept at the International Bureau of Weights and Measures. The definition of the base unit of length (the meter) has changed several times and is currently defined as the distance traveled by light in a vacuum during a time interval of 1/299,792,458 of a second. [IV.E]

111. b; The average run length is the number of cycles, time periods, or samples that elapse before the process signals out of control. When the process is out of control, the average run length should be small so that changes in the process are detected quickly. If the process is out of control, operators should adjust the process. If an operator does not adjust the process, this underadjustment can introduce additional variation into the process. [VI.F.6]

112. c; A contour plot of the fitted model can help find optimal settings of factors. A contour plot displays the predicted response over the range of the significant factors. Note that main effects plots and interaction plots can also provide useful information in selecting optimal settings of factors. [VI.H.4]

113. b; Since the underlying distribution is normal, the sampling distribution of $\overline{X}$ also follows a normal distribution with mean $\mu = \mu_{\overline{X}}$ and standard deviation

$$\sigma_{\overline{X}} = \sigma / \sqrt{n}.$$

Note that the large sample size requirement is not necessary here, since the filling process has a normal distribution. In this population, $\mu = 24$, $\sigma = 0.05$, and $n = 9$. Therefore, $\mu_{\overline{X}} = 24$ and

$$\sigma_{\overline{X}} = \sigma / \sqrt{n} = 0.05 / \sqrt{9} = 0.01667.$$

The probability of interest is

$$P(\overline{X} < 23.96) = P\left(\frac{\overline{X} - \mu_{\overline{X}}}{\sigma_{\overline{X}}} < \frac{23.96 - 24}{0.05 / \sqrt{9}}\right) = P(Z < -2.40)$$
$$= 0.0082.$$

[VI.C.2]

114. d; SPC (1) shows when a process is in statistical control (that is, the process variation appears to be random); (2) shows when a process is out of statistical control (that is, the process has nonrandom variability); and (3) can help determine whether or not a process is improving. Control charts indicate that the process is out of control, but they cannot identify what caused the out-of-control behavior. It is necessary to identify and correct out-of-control behavior when it is indicated in a control chart. [VI.F.1]

115. b; Consider the bias of the part at each time period. Recall that Bias = Observed average – Reference value.

Time	Day 1	1 month	2 months
Reference value	5.00	5.00	5.00
Observed average	5.007	4.975	4.824
Bias	0.007	–0.025	–0.176

Stability is a measure of how well the measurement system performs over time. In this example, the bias worsens as time elapses, indicating evidence of instability in the measurement system. *Linearity* measures how changes in the size of the part being measured will affect the bias of the measurement system. *Repeatability* is the variability of the test instrument when used to measure the same part. *Reproducibility* is the variability due to different operators or setups of the measurement system measuring the same part. [IV.F]

116. a; As stated in *The ASQ Certified Quality Engineer Handbook*, Fifth Edition, Seiichi Nakajima noted the ideal OEE condition of >90% availability, >95% efficiency, and >99% quality based on his experience. The ideal OEE is therefore documented by Nakajima as greater than or equal to 85%. [V.D]

117. d; The first region of the bathtub curve is known as the "early life" region and is characterized by a decreasing failure rate function. Since the failure rate is time-dependent in this region, the Weibull distribution is frequently used to model this region of the bathtub curve. [III.E.3]

118. d; The diagram displays a parallel system with three components. The reliability of a parallel system is

$$R_s(t) = 1 - [F_1(t) \times F_2(t) \times F_3(t)] = 1 - [(1 - 0.99) \times (1 - 0.92) \times (1 - 0.95)]$$
$$= 1 - [0.01 \times 0.08 \times 0.05] = 1 - 0.00004 = 0.99996.$$

[III.E.2]

119. a; This is an example of error-proofing, also known as *poka-yoke*. The order will not be processed unless there is complete shipping information and the address has been verified as valid. These extra steps should help reduce errors in shipping. [V.F]

120. b; A *statistic* is a characteristic of a sample and is used to estimate a population parameter. A *critical value* is used in hypothesis testing and is the value to which the value of the test statistic is compared to determine whether the null hypothesis can be rejected. [VI.B.1]

121. a; After risk undergoes treatment, the remaining risk is residual risk. Inherent risk is the raw, untreatable risk. Secondary risks are those arising from a treatment. Material risks are specific to the material used. [A.1]

122. d; A fractional factorial experiment uses fewer runs; however, not all interactions of interest may be estimable separately from main effects or other interactions. Some of the effects may be confounded with other effects. This can make it difficult to identify which factor or interaction is significant. [VI.H.5]

123. c; If the failure rate is decreasing, constant (such as when the reliability of the system has an exponential distribution), or unknown, maintaining the system in an attempt to prevent failures will not affect the probability that the system will fail in the next instant. Therefore, a preventive maintenance policy is not recommended. [III.E.2]

124. a; The *power* of a hypothesis test is defined as the probability of correctly rejecting a false null hypothesis:

$$\text{Power} = 1 - \beta = 1 - P(\text{Type II error}).$$

Type II error is defined as the probability of failing to reject H_0 when in fact H_0 is false. Type I error is defined as the probability of rejecting H_0 when in fact H_0 is true. [VI.D.2]

125. b; The scatterplot shows that there is not a linear relationship between the two variables. The correlation coefficient is 0. Note that the correlation coefficient can only have a value between –1 and 1. [VI.E.2]

126. a; Given that the lot size is 50,000, the sample size code letter from the ANSI/ASQ Z1.4-2003 sampling tables for normal, level II sampling is N. Using ANSI/ASQ Z1.4-2003 Table II-A, the required sample size is 500. Using the AQL of 1.0%, the acceptance number is ten and the rejection number is eleven. [IV.C.2]

127. d; The necessary conditions for a random variable to follow the binomial distribution are (1) a fixed number of trials, (2) the trials are independent, (3) each trial results in one of two possible outcomes, and (4) the probability of success is constant from trial to trial. In this case, $n = 50$, the parts are independent, each part can only be classified as defective or non-defective, and the probability of a defective part is assumed to be 0.01. Therefore, the number of defective parts will form a binomial distribution. [VI.C.2]

128. b; Let X represent the diameter of the steel rod. The upper specification limit is 1.58 and the lower specification limit is 1.42. First consider the proportion of steel rods that are within the specification limits, which can be written as $P(1.42 < X < 1.58)$. Since the diameter of the steel rod is normally distributed and the process is stable, the standard normal distribution can be used to determine this fraction.

From the control charts, an estimate for μ is $\hat{\mu} = \bar{\bar{X}} = 1.4983$. The point estimate for the process standard deviation is

$$\hat{\sigma} = \frac{\bar{R}}{d_2} = \frac{0.0906}{2.059} = 0.044.$$

The value for d_2 can be found in Appendix B of *The ASQ Certified Quality Engineer Handbook*, Fifth Edition for $n = 4$.

The fraction of steel rods that meet the specification limits is

$$P(1.42 < X < 1.58) = P\left(\frac{1.42 - 1.4983}{0.044} < \frac{X - \bar{\bar{x}}}{\hat{\sigma}} < \frac{1.58 - 1.4983}{0.044}\right) = P(-1.78 < Z < 1.86)$$

$$= P(Z < 1.86) - P(Z < -1.78) = 0.9686 - 0.0375 = 0.9311.$$

Therefore, the fraction of steel rods that are outside the specifications is $1 - 0.9311 = 0.0689$.

Approximately 6.89% of steel rods will fall outside the specification limits. [VI.G.2]

129. d; A tree diagram can take on several shapes. A *fault tree* is used to discover a hierarchical relationship between events that led to a failure in a process. Similarly, an *event tree* defines the relationship between events that lead to a given outcome. A *concept fan* is used during the creative process of finding alternate concepts that achieve the same requirements as a process or product. [V.B]

130. c; When using a double sampling plan, after the first sample, the lot is accepted, rejected, or another sample is drawn. If the number of defective items in the first sample, d_1, is less than or equal to the acceptance number, then the lot is accepted. If d_1 is greater than the rejection number, then the lot is rejected. If d_1 is more than the acceptance number but less than the rejection number, then another sample is drawn. In this case, $d_1 = 6$ and $\text{Ac} < d_1 < \text{Re}$. Therefore, another sample must be drawn before a decision is made. [IV.C.2]

131. b; Confirming is the only option that would be considered active listening. Listening to music and simultaneous speaking hinder communication. Being defensive on a team will quickly defeat the purpose of having a team. It can be overcome with a good facilitator and team leader, as well as a team willing to work together for a higher purpose. [I.F]

132. b; p and np control charts are based on the binomial distribution. [VI.F.5]

133. c; Let N be the event that a nonconforming part is made and let D be the event that a part is made during the day. The given condition is that the part is nonconforming, so

$$P(D|N) = \frac{P(D \cap N)}{P(N)} = \frac{5/284}{38/284} = 0.13.$$

[VI.B.3]

134. c;

$$\bar{x} = \frac{\Sigma_{i=1}^{n} x_i}{n} = \frac{0.183+0.190+0.180+0.189+0.190+0.188+0.186+0.185}{8} = 0.186$$

$$s = \sqrt{\frac{\Sigma_{i=1}^{n}(x_i - \bar{x})^2}{n-1}} = \sqrt{\frac{(0.183-0.186)^2 + \ldots + (0.185-0.186)^2}{8-1}} = \sqrt{\frac{0.000091}{7}} = 0.0036$$

Since the population variance is unknown, and the underlying distribution is normally distributed, a 95% confidence interval with the t-distribution can be used. The formula for the 100(1 – α)% confidence interval on the population mean μ is

$$\bar{x} - t_{\alpha/2,k}\frac{s}{\sqrt{n}} \le \mu \le \bar{x} + t_{\alpha/2,k}\frac{s}{\sqrt{n}}$$

$$\bar{x} = 0.186,\ s = 0.0036,\ \alpha = 0.05,\ n = 8,\ k = n-1 = 7,\ t_{0.05/2,7} = t_{0.025,7} = 2.365$$

$$0.186 - 2.365\left(\frac{0.0036}{\sqrt{8}}\right) \le \mu \le 0.186 + 2.365\left(\frac{0.0036}{\sqrt{8}}\right)$$

$$0.183 \le \mu \le 0.189.$$

[VI.D.1]

135. b; An independent contractor not associated with the auditee would be considered an external quality auditor. [II.D.2]

136. a; Histograms are used to understand the shape, spread, and distribution of a set of data. They can, therefore, be useful to compare the data against a target value or specification limits. Scatter diagrams are used to depict the relationship between two continuous variables, Pareto charts are used to identify the most common defects or occurrences in a process, and control charts are used to identify out-of-control behavior in a process. [V.A]

137. d; Let D_1 denote the probability that the lot is accepted or rejected on the first sample. Let d_1 denote the number of defects found in the first sample. The lot is accepted if there are two or fewer defects noted in the first sample. The lot is rejected if there are five or more defects in the first sample. (Another sample will be drawn if three or four defects are found in the first sample). Therefore, $P(D_1) = P(d_1 \leq 2) + P(d_1 \geq 5)$.

$$P(d_1 \leq 2) = \sum_{d_1=0}^{2} \binom{n_1}{d_1} p^{d_1}(1-p)^{n_1-d_1} = \sum_{d_1=0}^{2} \binom{75}{d_1} 0.05^{d_1}(1-0.95)^{75-d_1} = 0.2697$$

$$P(d_1 \leq 5) = 1 - P(d_1 \leq 4) = \sum_{d_1=0}^{4} \binom{75}{d_1} 0.05^{d_1}(1-0.95)^{75-d_1} = 1 - 00678 = 0.3211$$

Therefore, $P(D_1) = 0.2697 + 0.3211 = 0.5908$. [IV.C.2]

138. c;

$$\text{MTTF} = 115 = \frac{1}{\lambda} = \frac{1}{\text{Failure rate}}.$$

This means that the failure rate,

$$\lambda = \frac{1}{115}.$$

Since the reliability of the dishwasher has an exponential distribution, the probability that a randomly selected dishwasher survives past t months is $R(t) = e^{-\lambda t}$. Therefore, the probability that a randomly selected dishwasher fails by 115 months is

$$F(115) = 1 - R(115) = 1 - e^{-\frac{115}{115}} = 1 - e^{-1} = 1 - 0.3679 = 0.6321.$$

[III.E.2]

139. c; The ANOVA table can be used to test for significance of regression. The null and alternative hypotheses are $H_0 : \beta_1 = 0$ versus $H_a : \beta_1 \neq 0$. Note that the ANOVA approach in regression analysis only tests the slope of the regression equation. There are two approaches to test for significance. If the test statistic $F_0 > F_{\alpha,1,n-2} = F_{0.05,1,14} = 4.60$, H_0 would be rejected. Since $F_0 = 52.53 > 4.60$, H_0 is rejected, and it is concluded that there is a statistically significant linear relationship between the age of the car and its cost. Note also that the p-value is 0, which is less than the significance level $\alpha = 0.05$. The null hypothesis is again rejected. [VI.E.1]

140. a; From the information provided, $\bar{u} = 0.214$, $n = 28$, $m = 50$.

$$\text{LCL} = \bar{u} - 3\sqrt{\frac{\bar{u}}{n}} = 0.214 - 3\sqrt{\frac{0.214}{28}} = -0.048 \rightarrow 0.$$

Note that the formula gives a negative value for the lower control limit. By convention, the LCL is rounded to 0. [VI.F.5]

141. c; The main benefit of an FMEA is that it helps teams to identify and eliminate/reduce the negative effects of potential failures before they occur. However, FMEA is expensive. Note that FMEA can be used as a stand-alone tool and does not need to be part of a larger quality system such as Six Sigma. [III.E.4]

142. d; *Muda* is another name for waste. *Kaizen* is a performance improvement methodology that emphasizes incremental changes to improve a system. *Kanban* is a visual system that helps to manage the procedure for resupplying tools, components, and so on, in the workplace. *Seiton* is one of the five S's and means proper arrangement. [V.D]

143. c; 100% sampling is appropriate when the fraction of nonconforming products is high. If acceptance sampling were used in this scenario, then most of the lots under acceptance sampling would be rejected, and those lots that are accepted would be accepted because of sampling variability. [IV.C.1]

144. a; The process data in a capability study *must* come from a normal distribution. If the data are not normally distributed, the capability indices are not valid. A histogram and/or normal probability plot can be used to assess whether the process data come from a normal distribution. If the data are nonnormal, a transformation to induce normality may be required. [VI.G.1]

145. b; Selecting the range of factors in an experimental design is extremely important. In this problem, the range of the factor has been decreased. If the range of factors is too narrow, it is possible to miss important effects. [VI.H.2]

146. b; The histogram shows that the data set has a few extremely large values, indicating that the distribution is right-skewed. Left-skewed distributions have unusually small observations. A symmetric, or bell-shaped, distribution has no apparent outliers or unusual observations. A bimodal distribution has two peaks of frequent observations. [VI.A.6]

147. c; As the degrees of freedom approach infinity, the *t*-distribution becomes the standard normal distribution. [VI.C.1]

148. d; Regulatory requirements, customer needs, product or process capability, and product reliability are some of the inputs that will place requirements on the design. [III.B.3]

149. b; The employees have made small, incremental improvements to their processes at the pharmacy. These actions describe *kaizen* activities—incremental changes that lead to small improvements, but improve their process. The changes they have made were simple and did not require expensive technology or equipment, one of the features of kaizen improvement. [V.C]

150. c; The driver in a Critical to Quality Tree is what is needed to achieve what the customer asks for and will have lower-level Critical to Qualities that define what is needed to accomplish the driver. [III.B.1]

151. c; Of the options, only the frequency of the risk tracking review and reporting can be stated at the time of risk management planning. [B.1]

152. a; Variables sampling plans use the actual measurements of the product of interest. Therefore, these sampling plans determine how close the process is performing relative to specification limits. Attributes sampling plans only tell us whether to accept or reject a lot and do not provide any additional information about the process. [IV.C.2]

153. b; The parameter of interest is the difference in average fill volume, $\mu_1 - \mu_2$. The null and alternative hypotheses are $H_0 : \mu_1 - \mu_2 = 0$ versus $H_a : \mu_1 - \mu_2 \neq 0$. The significance level is $\alpha = 0.10$. This is a two-tailed test, so the null hypothesis is rejected if the test statistic is less than $-z_{\alpha/2}$ or greater than $z_{\alpha/2} = z_{0.05} = 1.645$. The test statistic is

$$z_0 = \frac{(\bar{x}_1 - \bar{x}_2) - \Delta_0}{\sqrt{\dfrac{\sigma_1^2}{n_1} + \dfrac{\sigma_2^2}{n_2}}} = \frac{(15.95 - 16.02) - 0}{\sqrt{\dfrac{0.05^2}{24} + \dfrac{0.04^2}{25}}} = -5.398.$$

Since $-5.398 < -1.645$, the null hypothesis is rejected. At the 10% significance level, there is a significant difference in average fill volume between the two methods. [VI.D.2]

154. d; Single minute exchange of die (SMED) is a method for reducing changeover times of machines. It has been attributed to Shigeo Shingo and his Toyota Production System. The goal of implementing SMED is to reduce the time required to switch a machine or assembly to a different product. [V.D]

155. c;

$$\lambda = \text{Failure rate} = \frac{\#\ \text{failures}}{\text{Total test time}} = \frac{62}{350 \times 8} = 0.0221.$$

Therefore,

$$\text{MTBF} = \frac{1}{\lambda} = \frac{1}{0.0221} = 45.25.$$

Since the mean time to repair the components is three hours, the steady-state availability of the component is

$$A = \frac{\text{MTBF}}{\text{MTBF} + \text{MTTR}} = \frac{45.25}{45.25 + 3} = 0.9378.$$

[III.E.2]

156. a; Continuous data can take any value over an interval of numbers. Age can be measured on a continuous scale. Gender and hair color are not numeric data. Number of members in the household is discrete data and can only take on certain values. For example, there cannot be 2.5 members in a household. [VI.A.1]

157. a; The normality assumption can be assessed from a normal probability plot of the residuals. They should form a straight line in the plot. If the normality assumption is not valid, then a transformation of the response variable may be necessary to perform the analysis. [VI.H.4]

158. c; The second region of the bathtub curve is the constant failure rate region. Since failures occur randomly over time in this region, the failure rate is constant. The exponential distribution is frequently used to model the reliability in this region. [III.E.3]

159. c; The variance can be calculated from the following equations.

$$E(X) = \mu_X = \Sigma x f(x) = 0(0.86) + 1(0.06) + 2(0.04) + 3(0.03) + 4(0.01) = 0.27$$

$$\begin{aligned} V(X) &= \Sigma(x-\mu)^2 f(x) \\ &= (0-0.27)^2(0.86) + (1-0.27)^2(0.06) + (2-0.27)^2(0.04) + (3-0.27)^2(0.03) \\ &\quad + (4-0.27)^2(0.01) = 0.5771 \end{aligned}$$

[VI.C.2]

160. d; Any problem identified must provide an opportunity for improvement. Therefore, to qualify as a performance improvement project, the problem must have variable performance from the established standard, be a deviation from the facts or perception, and its cause must be unknown. If what caused the problem is known, it's not a problem. [V.E]

161. d; Violations of the ASQ Code of Ethics include: (1) Accurately represent and do not mislead others regarding professional qualifications, including education, titles, affiliations, and certifications; and (2) Strive to uphold and advance the integrity, honor, and dignity of the Quality profession. [1.C]

162. c; Resolution IV designs have main effects confounded with three-factor or higher interactions, and two-factor interactions confounded with two-factor or higher interactions. With a resolution IV design, it is unknown for sure that a two-factor interaction is truly significant or whether the two-factor interaction it is aliased with is significant. Additional experimental runs may be needed to break these aliases. [VI.H.5]

163. d; The population does not have natural groups; therefore, stratified sampling is not necessary for this project. Simple random sampling will allow each customer record to have an equal probability of being selected as part of the sample. Double sampling and sequential sampling are methods used for acceptance sampling and are not appropriate here. [VI.A.4]

164. b; The assumptions for the least squares approach to linear regression are that the errors e_i are independent, normally distributed with mean 0, and have constant variance σ^2. Note that these assumptions can be verified using residual analysis. [VI.E.1]

165. a; From the information provided, $\mu = 97$, $\sigma^2 = 85$, $n = 24$. The standard error is defined as

$$s.e.(\bar{X}) = \sigma / \sqrt{n} = \sqrt{85} / \sqrt{24} = 1.88.$$

[VI.D.1]

166. a; According to the *ASQ Certified Quality Engineer Handbook*, Fifth Edition, the following table provides a reference guide in identifying a risk-based audit approach to process auditing throughout the quality management system.

Design Change	Risk-based approach will drive: • **Level of evaluation/testing (e.g., human factors/usability, verification, validation, clinical investigations)** • Regulatory requirements (e.g., notifications to Notified Body, Competent Authority, and regulatory filing/registrations) • Level of change approval (e.g., Engineer, Manager)

167. a; The floors and room signs match in color to the designated area of the hospital. This is an example of visual control—the added visual cues help patients and employees know where they are in a complicated building like a hospital. [V.D]

168. a; *c* and *u* control charts are based on the Poisson distribution. [VI.F.5]

169. c; During data analysis, data that are collected automatically and/or manually need to be cleaned, from unusual observations as well issue of missing data needs to be addressed. Certified Quality Engineers need to be aware that some missing data may be coded as zero values. In addition, before analysis is performed, additional tests such as data normality are required.

170. c; Let X represent the dimension of the component. The upper specification limit is 13.30 and the lower specification limit is 8.30. Of interest is the fraction within the specification limits. This can be written as $P(8.30 < X < 13.30)$. Since X is normally distributed, the standard normal distribution can be used to determine this fraction.

$$\begin{aligned}
P(LSL < X < USL) &= P(8.30 < X < 13.30) \\
&= P\left(\frac{8.30 - 10.45}{1.05} < \frac{X - \bar{x}}{\hat{\sigma}} < \frac{13.30 - 10.45}{1.05}\right) \\
&= P(-2.05 < Z < 2.71) = P(Z < 2.71) - P(Z < -2.05) \\
&= 0.9966 - 0.0202 = 0.9764.
\end{aligned}$$

Therefore, approximately 97.64% of components will fall within specification for this dimension. [VI.G.3]

171. d; Warranty costs would fall under the umbrella of external failure costs. [II.E]

172. c; From the ANOVA table, the *p*-values can be used to determine which factors have a significant effect on the response variable *y*. If the *p*-value $< \alpha$, then the null hypothesis is rejected. If the *p*-value $> \alpha$, then the null hypothesis is not rejected. From this ANOVA table, the *p*-value for factor A is less than $\alpha = 0.05$. Therefore, factor A has a significant effect on the response *y*. The *p*-values for factor B and the interaction effect between factors A and B are both greater than α. Therefore, they do not have a significant effect on the response variable *y*. [VI.D.5]

173. b; The takt time for this system is $10 \times 60 \times 60 \div 750 = 48$ seconds. The supplier must average one item every forty-eight seconds. Because the cycle time (forty seconds) is less than the takt time, the supplier is likely to meet the customer demand without changing their current production. [V.D]

174. b; The resolution of a design is equal to the length of the smallest word in the defining relation. Since the smallest word is four characters long in this defining relation, the design is resolution IV. [VI.H.5]

175. c; The correlation coefficient can be calculated from the following equations.

	x	y	xy	x^2	y^2
	5.2	26.7	138.84	27.04	712.89
	6.1	27.5	167.75	37.21	756.25
	3.2	24.9	79.68	10.24	620.01
	4.6	25.5	117.3	21.16	650.25
Σ	19.1	104.60	503.57	95.65	2739.40
Mean	04.775	26.15			

$$S_{xx} = \sum_{i=1}^{4} x_i^2 - \frac{\left(\Sigma_{i=1}^4 x_i\right)^2}{n} = 95.65 - \frac{(19.1)^2}{4} = 4.4475$$

$$S_{xy} = \sum_{i=1}^{4} x_i y_i - \frac{\left(\Sigma_{i=1}^4 x_i\right)\left(\Sigma_{i=1}^4 y_i\right)}{n} = 503.57 = \frac{(19.1)(104.6)}{4} = 4.105$$

$$S_{yy} = \sum_{i=1}^{4} y_i^2 - \frac{\left(\Sigma_{i=1}^4 y_i\right)^2}{n} = 2739.40 - \frac{(26.15)^2}{4} = 4.11$$

$$r = \frac{S_{xy}}{\sqrt{S_{xx}S_{yy}}} = \frac{4.105}{\sqrt{(4.4475)(15.024)}} = 0.96$$

Because $r = 0.96$, there is a strong linear association between the two variables. [VI.E.2]

176. b; In the *forming* phase, an important aspect is developing the mission. [I.D]

177. c; Since MTBF = 8, the failure rate $\lambda = 1/8$. One interpretation of the failure function $F(t)$ is the fraction of all rechargeable batteries that fail by time t.

$$F(10) = 1 - R(10) = 1 - e^{-\lambda(10)} = 1 - e^{-\frac{10}{8}} = 1 - 0.2865 = 0.7135.$$

Approximately 71% of all rechargeable batteries will fail by ten hours. [III.E.2]

178. c; The histogram shows that there are at least two extreme observations in this data set. Therefore, it is most appropriate to use a resistant measure such as the median. The median is not heavily influenced by extreme observations, while the mean is. The standard deviation is a measure of variability/spread, not central tendency. The standard deviation is also not a resistant measure. [VI.A.6]

179. b; Process decision program charts are primarily used to help evaluate a process implementation. They allow a team to assess the implementation plan, identify potential deviations, or develop contingency plans. Activity network diagrams are primarily used to help manage the schedule and timing of a project or set of actions. [V.B]

180. b; Radio-frequency identification (RFID) is an effective method for identification and tracking. In healthcare, RFID has been used effectively in patient care by tracking patient movements while they are hospitalized. [IV.B.1]

181. b; Modern training methods were necessary to present new methods to the workforce. [I.A.1]

182. c; Rational subgrouping involves selecting samples such that if there is special cause variation present in the process, there will be a greater probability of variation between samples, while the variation within samples is small. Therefore, the variation within samples is minimized, and the variation between samples is maximized. The selection of samples is important when constructing control charts to detect out-of-control behavior in the process. [VI.F.4]

183. a; Staying with the original perspective experts and customers have when it comes to value (a positive outcome in this case), then quality would be the correct answer. [I.G]

184. d; From the ANOVA table, none of the factors in this experiment has a significant effect on the response. The *p*-values for all main effects and interaction terms are relatively large (> 0.10); therefore, the conclusion is that none of the factors significantly affect the response. [VI.H.4]

185. c; A one sample *t*-test is appropriate where $H_0 : \mu = 18$ and $H_a : \mu \neq 18$. The degrees of freedom for a one-sample *t*-test are $k = n - 1 = 24 - 1 = 23$. Since $\alpha = 0.05$, H_0 is rejected if $t_0 < -t_{0.025,23}$ or $t_0 > t_{0.025,23}$. Therefore, the critical values for this hypothesis test (found in a *t*-table) are ±2.069. Note that the test statistic for this hypothesis test is

$$t_0 = \frac{\bar{x} - \mu_0}{s / \sqrt{n}} = \frac{17.98 - 18}{0.03 / \sqrt{24}} = -3.266.$$

Since $-3.266 < -2.069$, H_0 is rejected, and the conclusion is that the mean fill volume of soda is not equal to 18 oz. [VI.D.2]

186. a; This is an example of a fail-safe device. Since the components can only be attached to the tool in one way, the tool has been error-proofed from incorrect assembly. [V.F]

187. a; Note that the data are paired since the same component was measured twice, once by each inspector. Therefore, the confidence interval for paired data must be used.

Component	1	2	3	4	5
Inspector A	10.2	9.8	10.1	10.3	10.4
Inspector B	10.0	9.9	10.3	10.0	10.3
$d_i = A_i - B_i$	0.2	–0.1	–0.2	0.3	0.1

Let μ_1 represent the mean length measured by inspector A and let μ_2 represent the mean length measured by inspector B. Let μ_D represent the true mean difference between the two populations. $\mu_D = \mu_1 - \mu_2$.

The necessary summary statistics are:

$$\bar{d} = \frac{\Sigma_{i=1}^{n} d_i}{n} = \frac{0.2+(0.1)+(-0.2)+(0.3)+(0.1)}{5} = 0.06$$

$$s_d = \sqrt{\frac{\Sigma_{i=1}^{n}\left(d_i - \bar{d}\right)^2}{n-1}} = \sqrt{\frac{(0.2-0.06)^2+\ldots+(0.10-0.06)^2}{5-1}} = 0.2074$$

A 100(1 – α)% two-sided confidence interval on the parameter μ_D is given by

$$\bar{d} - t_{(\alpha/2,k)}\left(\frac{s_d}{\sqrt{n}}\right) \le \mu_D \le \bar{d} + t_{(\alpha/2,k)}\left(\frac{s_d}{\sqrt{n}}\right)$$

$$\alpha = 0.05,\ n = 5,\ k = n-1 = 4,\ t_{(\alpha/2,k)} = t_{0.025,4} = 2.776$$

$$0.06 - 2.776\left(\frac{0.2074}{\sqrt{5}}\right) \le \mu_D \le 0.06 + 2.776\left(\frac{0.2074}{\sqrt{5}}\right)$$

$$-0.197 \le \mu_D \le 0.317.$$

There is 95% confidence that the true mean difference between the length of the components measured by inspectors A and B is between –0.197 and 0.317. Note that 0 is contained in the confidence interval, indicating that the null hypothesis is not rejected ($H_0 : \mu_D = 0$). [VI.D.3]

188. b; Let T represent the time between arrivals at the store. T has an exponential distribution with mean $\mu = 4$ minutes. Since

$$\mu = \frac{1}{\lambda},\ \lambda - \frac{1}{\mu} = \frac{1}{4} = 0.25.$$

The cumulative density function for an exponentially distributed random variable T is given by

$$F(x) = P(T \le x) = 1 - e^{-\lambda x}.$$

Therefore,

$$P(T > 5) = 1 - F(5) = 1 - P(T \le 5) = 1 - (1 - e^{-\lambda \times 5}) = e^{-0.25 \times 5} = 0.2865.$$

[VI.C.1]

189. c; The number of degrees of freedom associated with the numerator in the F-test is equal to $a - 1$, where a is the number of levels of factor A. Therefore, since the degrees of freedom is 5, there are six levels of factor A. Similarly, there are five levels of the blocking factor. [VI.H.4]

190. d; This is an example of waste due to lost creativity. The management at the factory should encourage a quality team to investigate the proposed idea. This could lead to process improvements and will also increase employee morale as they contribute to performance improvement efforts. [V.D]

191. d; The sample average is

$$\overline{x} = \frac{\Sigma_{i=1}^{n} x_i}{n} = \frac{5.3+8.6+7.2+9.1+7.5+3.5}{6} = 6.8667.$$

The sample variance is

$$s^2 = \frac{\Sigma_{i=1}^{n}(x_i - \overline{x})^2}{n-1} = \frac{(5.3-6.8667)^2 + \ldots + (3.5-6.8667)^2}{6-1} = \frac{22.2933}{5} = 4.459.$$

[VI.A.6]

192. c; The control limits for the MR control chart are defined as UCL = $D_4\overline{MR}$ and LCL = $D_3\overline{MR}$. The constants D_3 and D_4 can be found using Appendix C of *The ASQ Certified Quality Engineer Handbook*, Fifth Edition for $n = 2$.

$$UCL = (3.267)(1.75) = 5.72$$

$$LCL = (0)(1.75) = 0$$

Note that the control limits for the individuals chart are:

$$\text{UCL} = \bar{x} + 3\frac{\overline{MR}}{d_2} = 17.90 + 3\left(\frac{1.75}{1.128}\right) = 22.55$$

$$\text{LCL} = \bar{x} - 3\frac{\overline{MR}}{d_2} = 17.90 - 3\left(\frac{1.75}{1.128}\right) = 13.25$$

[VI.F.5]

193. a; Detecting autocorrelation can be done with many methods, including trend analysis or moving average smoothing. [VI.E.3]

194. c; Dull tools, large changes in raw materials, and a broken machine are all special causes of variation. They are not an inherent part of the process and can be controlled by the operators. Variation in ambient humidity, however, is a natural part of a process, and cannot be controlled by the process operators. [VI.F.2]

195. a; The fifth data point is above the upper control limit. One of the rules to detect out-of-control behavior is that one point is more than 3 < from the centerline on either side. Note that a control chart does not provide information on whether the process is operating within the specification limits. A process could be out of statistical control, but still operating within the specification limits. Similarly, a process could be in control, but not be within the specification limits. [VI.F.6]

196. a; The task of maintaining the risk level below an acceptable level is one purpose of risk control. [VII.C.1]

197. d; The test statistic for a hypothesis test for two population variances is:

$$F_0 = s_1^2 / s_2^2.$$

The test statistic follows an F-distribution with degrees of freedom $k_1 = n_1 - 1$ and $k_2 = n_2 - 1$. [VI.D.2]

198. a; The plan–do–check–act (PDCA) cycle, also called the Shewhart or Deming (PDSA) cycle, provides a four-step, iterative procedure for implementing process changes. Total quality management (TQM) is an organization-wide quality program that focuses on customers' involvement in determining quality, employees' obligation to quality, and strong leadership from upper management in quality improvement efforts. Theory of constraints (TOC) is a problem-solving methodology that focuses on the weakest part of a process, and reengineering involves redesigning a process to see large improvements in quality of a product. [V.C]

199. c; From the information provided,

$$USL = 16.75, LSL = 14.25, \mu = 16.50, \sigma = 0.10.$$

The process is not centered, so the actual capability metric, C_{pk}, applies.

$$C_{pk} = \min\left\{\frac{USL-\mu}{3\sigma}, \frac{\mu-LSL}{3\sigma}\right\} = \min\left\{\frac{16.75-16.50}{3(0.10)}, \frac{16.50-14.25}{3(0.10)}\right\}$$
$$= \min\{0.83, 7.5\} = 0.83.$$

Since the actual capability is less than 1, the process is not considered capable. [VI.G.3]

200. b; Costs associated with spare parts would be under the umbrella of appraisal costs. [II.E]

201. b; There is a need to distinguish the level of impact so that the consequences can be compared and evaluated. Quantification of risk also depends on the occurrence or probability values. Filling out a risk register is not an end goal. Monitoring of risk involves monitoring both consequence and probability or occurrence. [B.2]

202. a; The diagram displays a series system with three components. The reliability of a series system is the product of the reliabilities of its components. Therefore, the reliability of the system is

$$R_s(t) = R_1(t) \times R_2(t) \times R_3(t) = 0.99 \times 0.92 \times 0.95 = 0.8653.$$

[III.E.2]

203. b; B represents the secondary datum. [III.C]

204. b; While a corrective maintenance policy allows for the maximum run time between repairs, it is not the most efficient or cost-effective. If there is a catastrophic failure, the repair time may be extensive, with high costs. [III.E.2]

205. b; This is an example of waste due to defect correction. The most frequent cause for returns was because the company shipped the wrong item to the customer. The company therefore had to perform rework—process the return, find the correct item, and pay for shipping. [V.D]

206. d; Informed people are not those actually performing the project tasks, but are important to keep informed so they can assist with removing roadblocks and ensuring financial and resource support. [I.B.2.d]

207. c; A *k*-out-of-*n* system design requires that a system made up of *n* subsystems must have a minimum of *k* subsystems functioning to operate. In this case, since the machine will fail if four of six subsystems fail, the machine has a 3-out-of-6 system design, since at most three subsystems can fail before the machine fails. [III.E.2].

208. d; The experiment has $2^{8-2} = 64$ runs. Therefore, this experimental design is a one-fourth-fractional factorial design. [VI.H.5]

209. b; *Criticality* is an assessment of the consequences of a failure mode along with how frequently it occurs. [III.E.4]

210. b; *Internal activities* are defined as activities performed while the machine is down. *External activities* are defined as actions done in preparation for a machine changeover. Shingo recommended moving as many internal activities to external activities as possible. [V.D]

211. b; The mean and variance for a random variable that follows a Poisson distribution with rate λ are $\mu = \sigma^2 = \lambda$. Given that the standard deviation, $\sigma = 3.04$, $\sigma^2 = \mu = 3.04^2 = 9.24$. [VI.C.2]

212. b; International standards are necessary because they provide a consistent way for interpreting terms, vocabulary, and requirements regardless of language or location of the reader (different languages interpreting information differently was a key problem with standards from different countries and in different languages). [II.C]

213. a; The emphasis is on results; the award is not given for specific products or services, but to reward achievements in quality and performance excellence. [II.C]

214. d; Design review and field trials would fall under prevention costs. [II.E]

215. b; The FTA only uses binary data. This is often considered a limitation of the technique, although the method is very effective for failure analysis. [VII.B.2]

216. c; The risk priority number is a numerical result from the FMEA. The RPN is the result of multiplying three values: severity of failure, failure occurrence likelihood, and the likelihood of detecting the failure. [VII.B.2]

www.ingramcontent.com/pod-product-compliance
Lightning Source LLC
LaVergne TN
LVHW060030210926
848481LV00013B/3828
9781636941134